Modern Antenna

The modern antenna book series mainly covers the related antenna theories and technologies proposed and studied in recent years to solve the bottleneck problems faced by antennas, including binary coded antenna optimization method, artificial surface plasmon antenna, complex mirror current equivalent principle and low profile antenna, generalized pattern product principle and generalized antenna array, cross dielectric transmission antenna, metamaterial antenna, as well as new antenna technology and development. This series not only presents the important progress of modern antenna technology from different aspects, but also describes new theoretical methods, which can be used in modern and future wireless communication, radar detection, internet of things, wireless sensor networks and other systems.

The purpose of the modern antenna book series is to introduce new antenna concepts, new antenna theories, new antenna technologies and methods in recent years to antenna researchers and engineers for their study and reference. Each book in this series is thematic. It gives a comprehensive overview of the research methods and applications of a certain type of antenna, and specifically expounds the latest research progress and design methods. As a collection, the series provides valuable resources to a wide audience in academia, the engineering research community, industry and anyone else who are looking to expand their knowledge of antenna methods. In addition, modern antenna series is also open. More antenna researchers are welcome to publish their new research results in this series.

Xiang Gao · Chuanfei Ding · Wenfei Tang ·
Kun Guo

Conformal Array Antennas and Beamforming Technologies for Emergency Communication Applications

Springer

Xiang Gao
School of Cyberspace Science
and Technology
Beijing Institute of Technology
Beijing, China

Wenfei Tang
School of Information and Electronics
Beijing Institute of Technology
Beijing, China

Chuanfei Ding
School of Information and Electronics
Beijing Institute of Technology
Beijing, China

Kun Guo
School of Cyberspace Science
and Technology
Beijing Institute of Technology
Beijing, China

ISSN 2731-7986 ISSN 2731-7994 (electronic)
Modern Antenna
ISBN 978-981-95-2904-9 ISBN 978-981-95-2905-6 (eBook)
https://doi.org/10.1007/978-981-95-2905-6

This Springer imprint is published by the registered company Springer Nature Singapore Pte Ltd.
The registered company address is: 152 Beach Road, #21-01/04 Gateway East, Singapore 189721, Singapore

Contents

1 Introduction 1
1.1 The Antenna Requirements for Emergency Communication Applications 1
1.2 Antenna Classifications 4
1.2.1 Antenna Elements 4
1.2.2 Planar Array Antennas 6
1.2.3 Conformal Array Antennas and Beamforming Technologies 8
References 12

2 Balloon Conformal Array Antennas 19
2.1 A Balloon Conformal Log-Periodic Array Antenna 19
2.1.1 Comparative Analysis of Conformal Strip-Dipole and Bowtie-Dipole 19
2.1.2 Conformal Log-Periodic Balloon Array Antenna 22
2.1.3 Measurement of the Balloon Array Antenna 26
2.1.4 Balloon Conformal Array Antenna 31
2.2 A Balloon Conformal Array Antenna of Broadened Bandwidth, High Gain and Steerable Directional Beam Fed by a Flexible Phase Shifting Network 31
2.2.1 Design of Balloon-Based Conformal Array Antenna 32
2.2.2 Design of Flexible Phase Shifting Network 40
2.2.3 Experimental Measurements and Analysis 44
2.3 A Balloon Conformal Heterogeneous Array Antenna with Ultrawide Bandwidth and Half-Space Beam Coverage 51
2.3.1 Design and Analysis of the Balloon-Based Conformal Heterogeneous Array Antenna 52
2.3.2 Measurement, Simulation and Analysis 63
References 69

3 Conformal Array Antennas for Mobile Platforms 71
3.1 A Quad-Polarization Reconfigurable Conformal Array Antenna for Unmanned Aerial Vehicle Communication Applications 73
3.1.1 Antenna Design and Analysis 73
3.1.2 Simulation and Measurement 82
3.2 A Helmet-Based Hemispherical-Conformal Wearable Array Antenna with Broad Bandwidth, Wide-Angle Coverage and Reconfigurable Dual Circular Polarizations for WBAN Applications 90
3.2.1 Antenna Design and Analysis 90
3.2.2 Simulation and Measurement 104
References 116

4 DOA Estimation and Beamforming Algorithms Based on Conformal Sparse Arrays 117
4.1 Coherent Signal DOA Estimation Algorithm Based on Conformal Circular Arrays 117
4.1.1 One-Dimensional Coherent Signal DOA Estimation Algorithm of Conformal Circular Arrays Based on Sparse Reconstruction 118
4.1.2 Two-Dimensional Coherent Signal DOA Estimation Algorithm of Conformal Circular Arrays 130
4.2 Sparse Optimization Algorithm for Conformal Arrays 139
4.2.1 Principles of Array Radiation Pattern Synthesis 141
4.2.2 Optimization Model of the Sparse Cylindrically Conformal Array 143
4.2.3 Improved Snake Optimization Algorithm and Simulation Results 145
4.3 Interference Suppression Beamforming Algorithm for Conformal Sparse Array 154
4.3.1 Principles of Interference Suppression Beamforming 155
4.3.2 Interference Suppression Beamforming Algorithm for Sparse Circular Arrays 158
4.3.3 Simulation Results of Interference Suppression Beamforming for Conformal Sparse Arrays 164
References 171

5 Conclusion 173
5.1 Contributions of This Book 173
5.2 Future Work 175

Chapter 1
Introduction

1.1 The Antenna Requirements for Emergency Communication Applications

Natural disasters such as forest fires, earthquakes and mudslides, pose significant threats to human health and safety year by year. Emergency communication plays a vital role in combating these disasters and has garnered significant attention in the past decades [1]. These disaster-stricken areas typically have no communication capabilities: either occurring in remote, uninhabited mountainous-and-dense forests or having all local communication facilities destroyed by the disaster. For clarity, the mountainous forested terrain is taken as an example to elaborate the antenna requirements for emergency communication.

In dense forests and hilly regions, the channel condition is adverse (severe propagation attenuation and multipath effects), thus posing significant challenges to long-distance and stable emergency communication. For example, a large forest fire caused by lightning strike in Muli County of Sichuan Province, China in 2019. The conflagration zone is a mountainous area filled with deep valleys and dense trees at an altitude of about 3800 m. Poor geographical conditions hinder the construction of local communication facilities, which make the firefighters out of contact with each other thus causing huge casualties. Therefore, a robust emergency communication system is of critical importance to the mountainous and dense forest environment, which allows the front-line rescuers to timely know where there is a danger and how to protect their lives. It should be noted that since this book primarily focuses on array antennas, the subsequent analysis will examine their role in emergency communication systems specifically from an antenna perspective.

Figure 1.1 shows the simulated bit error ratio (BER) performance for different 5G multi-carrier schemes under different conditions. SUI-5 and SUI-6 channel models proposed by Stanford University to characterize mountainous forested environments, were later incorporated into IEEE 802.16 standardized channel models [2]. These BER performance metrics were obtained through simulations conducted

X. Gao et al., *Conformal Array Antennas and Beamforming Technologies for Emergency Communication Applications*, Modern Antenna,
https://doi.org/10.1007/978-981-95-2905-6_1

in the absence of channel coding, considering two distinct motion modes (low-velocity or high-velocity). As depicted by the transition from Fig. 1.1a–b (similarly observed from Fig. 1.1c–d), directional antennas achieve superior BER performance compared to omnidirectional antennas under identical conditions. This improvement is attributed to the directional antenna's narrow beam width, which mitigates multipath interference (weaker channel power of multipath).

Another crucial consideration for a communication link in the mountainous and dense forest environment is the transmission distance. Figure 1.2 shows the simulated BER versus the transmission distance for different multi-carrier communication schemes. Provided the transmitting antenna height is changed from 10 to 50 m, the maximum transmission distance is approximately doubled as depicted in Fig. 1.2a and b. Usually, it is difficult for the rescuers to place an antenna at that height in the mountainous and dense forest environment. Even so, a very satisfactory transmission

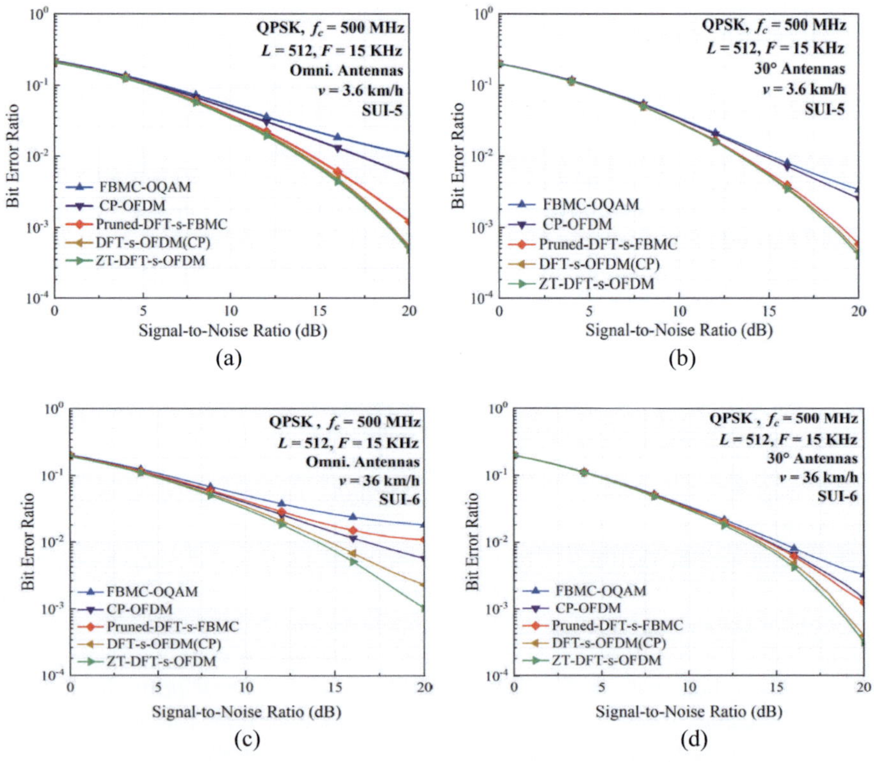

Fig. 1.1 Simulated BER performance for different multi-carrier schemes under different conditions. For a fair comparison, we set the carrier frequency f_c, the subcarrier spacing F and the subcarrier number L to constant values. **a** and **b** Are the simulated BER performance for QPSK modulations under SUI-5 channels using omnidirectional and directional antennas, respectively. **c** and **d** Are the simulated BER performance for QPSK modulations under SUI-6 channels using omnidirectional and directional antennas, respectively

distance is still not achieved. To solve this problem, an aerial station equipped with high-gain antennas acting as a transponder with the conformal directional antenna can be raised to the desired altitude for auxiliary communication.

Although high-gain antenna can significantly extend transmission range, their own narrow beam characteristics result in limited coverage. However, there are many users in the emergency communication scenario in the mountainous and dense-forest environment. These users include ground search and rescue personals, vehicles, low-altitude drones, etc., which may appear randomly in any position in the mountainous and dense-forest environment. Therefore, a wide beam coverage is required while maintaining a high gain. Furthermore, a wide bandwidth is favorable for supporting high-speed communication applications.

Emergency communication networks are composed of both fixed and mobile communication platforms. Conformal array antennas, characterized by high gain and wide coverage, can be applied to fixed platforms as communication hubs. However, mobile communication platforms during their operational deployment

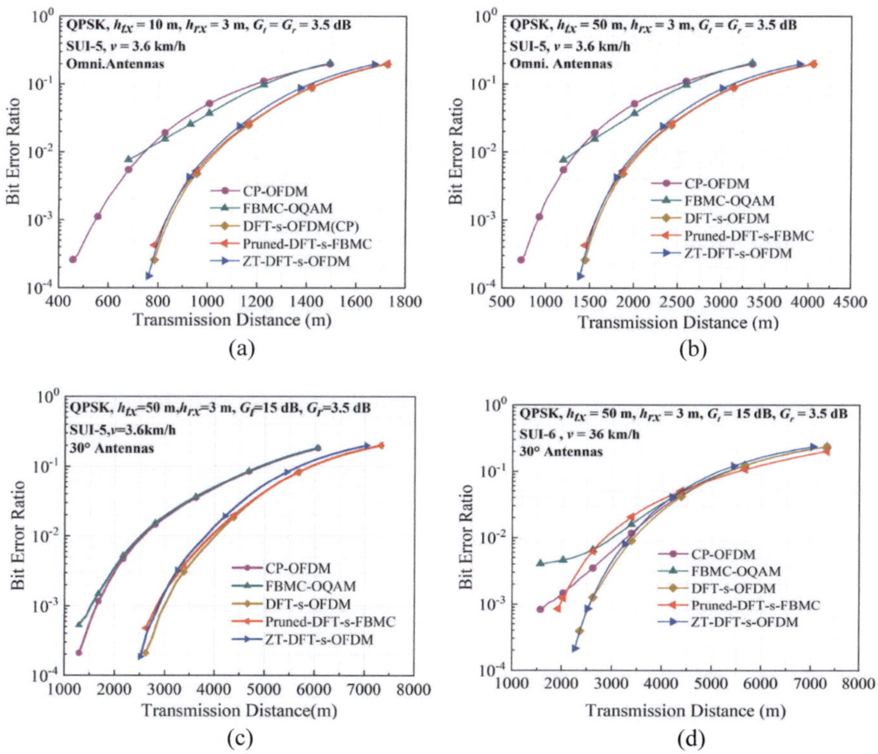

Fig. 1.2 Simulated link BER performance versus the transmission for the cases of: **a** and **b** are the BER performance when the transmitting antenna height is 10 m and 50 m, respectively. **c** The performance of the15 dB gain antenna. **d** Changes the SUI-6 sub-channel model and the speed to 36 km/h of the unmanned aerial vehicle (UAV)

are frequently subjected to attitude changes, dynamic environmental conditions, and electromagnetic interference from multiple temporary devices. These factors can readily induce severe multipath interference and signal blockage, challenges which conventional single-polarization antennas struggle to mitigate effectively. To address this, polarization reconfigurable conformal array antennas offer a solution by dynamically adjusting the polarization state of their main beam in real-time. This capability enables the suppression of multipath effects and facilitates the selection of an optimally polarization-matched propagation path, thereby significantly enhancing signal quality and anti-jamming capability. Consequently, conformal array antennas endowed with polarization reconfigurability constitute an essential enabling technology for establishing high-quality communication links on mobile platforms.

In harsh emergency communication applications, which are composed of both fixed and mobile carriers such as ballons and UAVs, there exist problems such as severe occlusion, severe path loss and complex multipath effects. In order to solve these problems, the research on conformal array signal processing technology, mainly including DOA estimation, sparse optimization and interference suppression beamforming algorithms, has become an important way to improve the performance of wireless communication.

In summary, the antenna requirements for emergency communication applications can be summarized below:

(1) High gain to achieve long distance communication.
(2) Wide bandwidth to achieve high-speed communication.
(3) Broad beam coverage to support users access from different locations.
(4) Polarization diversity for addressing polarization mismatch and multipath interference in mobile communication platforms.
(5) High-accuracy DOA estimation algorithm for determining the location information of the signal source in harsh emergency communication environments.
(6) High-performance sparse optimization and interference suppression beamforming algorithms for enhancing the flexibility and anti-interference ability.

1.2 Antenna Classifications

1.2.1 Antenna Elements

With the rapid development of modern communication technology, the antenna is the gateway for receiving and sending information, and its performance directly determines the upper limit of the communication system's capability. This book focuses on emergency communications, a special field that requires extreme reliability, flexibility and efficiency of communication systems, and deeply discusses the indispensable and central role of array antenna technology and its key advantages. However, to fully understand the value of array antennas, it is necessary to first examine the performance of antenna elements in current mainstream applications and the insurmountable gap in emergency scenarios [3].

1.2.1.1 Commonly Used Antenna Elements for Emergency Communications

The antenna element is composed of an independent radiation element and represents the most basic and common form in wireless communication systems. In modern terrestrial mobile communications and portable devices, several typical forms of antenna elements are widely used.

(1) Monopole antenna or whip antenna: Noted for its simple structure and low cost, it is widely used in portable radios, vehicle stations, some low-cost Internet of Things (IoT) devices and even some base station components. It is usually a quarter wavelength ($\lambda/4$) long and requires a ground plane for operation.
(2) Dipole antenna [4]: A classical omnidirectional antenna form, consisting of two segments of conductors with the length of approximately $\lambda/4$, does not need to rely on an external ground plane. It is often used in scenarios demanding slightly higher gain, such as amateur radio base stations, TV reception, etc.
(3) Microstrip antenna [5]: Its most notable advantages include thinness and ease of integration, as it can be easily combined with the manufacturing process of Printed Circuit Board (PCB). It is widely used in mobile phones, internal antennas of Wi-Fi routers, and various miniaturized devices and systems.
(4) Helix or spiral antenna: It can achieve a wide bandwidth and a reasonable gain within a compact volume, and is generally used in satellite communication terminals to provide quasi-omnidirectional coverage.
(5) Sleeve antenna: As a variant of monopole antenna, it is often used to improve bandwidth and increase gain, which is seen in some professional mobile communication devices and certain base station antenna elements.

1.2.1.2 Deficiencies of Antenna Elements in Emergency Communications

Emergency communication poses stringent demands on antenna systems that surpass those of conventional services. Antenna elements exhibit critical limitations in this context:

(1) Limitations in gain or coverage: During disasters, base stations may be destroyed or urgently require rapid coverage extension. Antenna elements are characterized by limited inherent gain and short propagation range. To cover broader areas or penetrate complex obstacles often necessitates increased transmit power or elevated antenna deployment, which is prohibitively expensive or infeasible in urgent scenarios. This difficulty makes it hard to establish long-distance backbone links, severely limits communication range between small portable

devices, and creates many communications blind spots across the search and rescue area.

(2) Weak resistance to multipath fading: A single element antenna is a single channel receiver and transmitter, and the disaster site environment is often harsh. The signal propagates among collapsed buildings and complex landforms, and the multipath fading is extremely serious. Reliance on a single signal path means communications are easily obstructed or subject to deep fading nulls. This results in a dramatically elevated risk of link failure and highly unstable received signal quality.

1.2.1.3 Array Antennas: The Key Out of Troubles

It is these fundamental bottlenecks faced by antenna elements in emergency communications that have spawned an urgent need for more advanced antenna technologies. Array antenna technology has the potential to fundamentally overcome these limitations as a means to spatially arrange multiple elements and precisely control the amplitude and phase of the feeding signals of each element to form the desired beam and radiation characteristics. Therefore, in-depth research and application of array antenna technology is a key breakthrough to improve the efficiency of emergency communication and establish a more reliable, flexible and large-capacity communication security system. The following chapters of this book will systematically describe the basic principles, design methods, key technologies of array antennas, and their specific applications and practices in various emergency communication systems and scenarios, providing readers with a comprehensive perspective and in-depth analysis of this future-oriented core technology of emergency communication.

1.2.2 Planar Array Antennas

To meet diverse application demands such as long-distance transmission, anti-interference capabilities, and cost-effectiveness in modern communication systems, antennas work as critical system components, which face increasingly stringent requirements. Compared to single-element antennas, planar array antennas are extensively deployed in diverse application scenarios due to their high gain and beam steering capabilities. Planar array antenna refers to an antenna system comprising identical radiating elements arranged periodically on a planar surface. In contrast to antenna arrays such as horn arrays, linear dipole arrays, and Vivaldi arrays, which utilize high-profile three-dimensional radiating structures, antenna elements in a planar array are configured in a coplanar arrangement. In recent years, various planar array antenna designs have been proposed, including back-cavity arrays, cavity array antennas, slot array antennas, and transmission-line-based planar arrays.

The back-cavity array antenna comprises back-cavity slot array antennas [3–8] and back-cavity patch array antennas [9–11]. Back-cavity array antennas are typically implemented by etching radiation slots on the upper surface of low-profile substrate integrated waveguide (SIW) cavities. The back-cavity patch array antenna incorporates a parasitic patch layer positioned above the cavity to enhance radiation and improve gain [9]. Distinguished from back-cavity array antennas, the cavity array antenna is free from metal patches serving as radiating enhancement structures and is not patterned with radiating slots on its top surface [12–14]. By directly utilizing the open-cavity structure as the radiating element, the cavity-backed array antenna effectively achieves a lower Q factor, thereby yielding wider operational bandwidths. Slot array antennas are also a type of planar array antenna composed of multiple slot radiating elements arranged in a specific geometric configuration. Electromagnetic radiation occurs when the slots are excited by electromagnetic fields provided by feed structures such as waveguides or microstrip lines [15–17]. Additionally, there exists a planar array antenna constructed based on planar transmission line structures [18–20]. This antenna alters the boundary conditions at the junction between the transmission line and free space by introducing discontinuities in the transmission line structure. Such discontinuities enable the guided modes propagating within the transmission line to couple with the propagation modes in free space, thereby converting bound waves into radiation energy.

Although cavity-backed array antennas, cavity array antennas, slot array antennas, and transmission line-based array antennas can offer high gain and meticulous beam control in specific scenarios, they generally struggle to adapt to the demands of emergency communications. Firstly, these antenna designs are typically structurally complex and bulky (e.g., cavity or cavity-backed elements requiring substantial enclosures), making them unsuitable for portability and rapid deployment. Secondly, slot array antennas are susceptible to performance degradation from moisture or dust intrusion, reducing their reliability under harsh emergency conditions. Thirdly, transmission line-based arrays rely on precise calibration and complex feeding networks, resulting in time-consuming deployment procedures and complex operation that hinder real-time response capabilities. Finally, the high cost of these antennas restricts their widespread use in resource-constrained emergency scenarios. Given the aforementioned limitations of planar arrays, they fail to meet the critical requirements of cost-effectiveness, high flexibility, and real-time responsiveness in emergency communication scenarios. Conformal array antenna integrated on surface of platform to maintain the aerodynamic integrity of the platform, which not only preserves the high-gain characteristics of a planar array but also achieves broader beam coverage.

1.2.3 Conformal Array Antennas and Beamforming Technologies

Fixed communication platforms and mobile communication platforms constitute essential components of emergency communication networks. In emergency communication scenarios, fixed platforms such as satellite communication vehicles and stratospheric balloons are typically deployed first to establish reliable, high-capacity backbone communication links. Following this, mobile platforms, including UAV systems or portable devices, are utilized to extend coverage into the disaster-affected area. Fixed platforms support the stable and efficient operation of the on-site command center, while mobile platforms provide communication support for rescue personnel. To construct a reliable, wide-coverage, and highly adaptive emergency communication network, conformal antennas suitable for both types of platforms, along with beamforming algorithms specifically designed for these conformal antennas, were investigated.

1.2.3.1 Conformal Array Antennas for Stationary Platforms

Forest fires, earthquakes, mudslides, and other natural disasters have occurred frequently in recent years. Emergency communication plays a vital role in combating these disasters and has garnered significant attention in the past decades; however, many disasters occur in dense forests and hilly regions with adverse channel conditions like severe propagation attenuation and multipath effects, thus posing significant challenges to high-quality emergency communications [1]. The use of airborne platforms like UAVs equipped with high-gain antennas can be a viable solutio [22, 23]; however, the UAVs feature relatively high cost and limited endurance restricted by battery capacity [24]. In contrast, inflatable balloons [25–28] have much lower cost, longer hang time, and smaller volume after deflation, which is, accordingly, more suitable for emergency communication applications in the hilly forest environment. Therefore, designing high-performance balloon conformal array antennas is critical for emergency communication applications.

Considering the advantage of good aerodynamic characteristics [29, 30], the conformal array antennas have been investigated in recent years. Currently reported ultrawideband conformal array antennas include the cavity-backed coupled array antenna [31], the tightly coupled dipole array [32] and the log-periodic array antenna [33]. These antennas are cylindrically conformal, but all of their profiles or weights are too high to be attached to the surface of a balloon, which would negatively affect the aerodynamic characteristics and floating capability of the balloon. A wideband and high-gain conformal transmitarray antenna [34] is also reported. Despite that the transmitarray itself features relatively low profile, the whole antenna system comprised of the transmitarray and the horn possess large physical space. Mounting such antenna into a balloon is impractical, since the horn feeder is relatively heavy and difficult to be located precisely at the required position inside the balloon. Similarly, the conformal array antennas [35–37] with their partial structures embedded into the carriers are not suitable for using as the balloon antennas. Due to their low-profile and light-weight properties, conformal thin microstrip array antennas [38–41] are more suitable for integrating into the balloon systems. These conformal microstrip arrays can realize beam steering, but they all require relatively complicated phase shifting network and suffer from narrowband operation, thus imposing severe restrictions on the balloon-based emergency communication applications. To date, some conformal balloon antennas are reported [42, 43], which exhibit satisfactory wideband operation performance. However, these balloon antennas are not array antennas, featuring very low gain and fixed beam characteristics, which can hardly accommodate the harsh hilly forest communication channel conditions associated with severe propagation attenuation and multipath effects. To the best of our knowledge, there has been few reports on the balloon-based conformal array antenna with relatively wide bandwidth, high gain, or broad beam coverage.

1.2.3.2 Conformal Array Antennas for Mobile Platforms

For mobile platforms such as aircraft and high-speed vehicles, conventional protruding antennas or planar array antennas can induce significant aerodynamic integrity. In contrast, conformal antennas, being integrated flush with the surface contour, offer superior aerodynamic performance [44–48]. By efficiently utilizing the available curved surface area, conformal antennas enable the implementation of array functionalities within spatially constrained systems [31, 33–35, 49–56].

A number of high-performance conformal array antennas have been proposed for various curved surfaces on mobile platforms. To achieve high-capacity communication for high-speed mobile platforms, a conformal low-profile log-periodic array mounted on a cylindrical conductor has been proposed [33]. The antenna shows an extremely wide impedance bandwidth up to 9.15:1 has been achieved with a very low profile of 0.053 λ_L. It can be potentially very useful for communication systems such as missiles or unmanned aerial vehicles. For geometrically complex surfaces constructed by integrating cylindrical and conical sections at the forebody of fixed-wing UAV or missiles, a full solid angle scanning cylindrical and-conical conformal phased array antenna is designed to achieve full solid angle scanning [35]. By combining three arrays, a conical conformal microstrip array antenna, a cylindrical conformal microstrip array antenna, and a cylindrical conformal Yagi array antenna, a full solid angle scanning has been achieved.

Mobile platforms such as aircraft and wearable devices undergo sustained dynamic attitude changes during motion. Simultaneously, the mobility of the platform will alter its spatial relationship with obstacles in the surrounding environment, leading to complex multipath effects. Under the dual influence of attitude changes and multipath channels, the polarization of signals received by antennas on mobile platforms undergoes non-negligible variations and fluctuations. In dynamic and complex environments, circular polarization and polarization-reconfigurable antennas offer advantages over single linear polarization (LP) antennas. Circularly polarized (CP) antennas can overcome polarization mismatch caused by attitude changes, suppress multipath interference, and mitigate signal fading [57–65]. Polarization-reconfigurable antennas enhance link reliability and data rates significantly by switching polarization states to achieve optimal signal reception intensity [66]. Additionally, they can replace multiple fixed antennas with singular functions, noticeably reducing system weight and volume, trimming costs, and lowering installation complexities. For high-speed mobile platforms such as UAVs, missiles, and vehicles, some multi-polarized conformal array antennas suitable for various curved surfaces have been proposed. For example, a wideband and wide-scanning cylindrically conformal phased array with a low profile and convenient assembly characteristics that can be totally attached to a cylindrical carrier surface is presented [32]. The antenna aperture is made of tightly coupled dipole elements. Besides the cylindrical dual-polarized conformal array antenna, a novel array arrangement and a dual-polarized mono-pulse comparator suitable for conformal array is developed [67]. With the advantages of dual-polarized mono-pulse and entirely conformal structure, it is a promising candidate as target detection and tracking antennas for various aircraft applications. Furthermore, regarding spherical surfaces commonly used in mobile platforms, a low-profile array on a doubly curved surface that supports wide angle electronic scanning and ultrawide bandwidth (7:1 ratio) is presented [68]. The array elements are based on the balanced antipodal Vivaldi antenna. The antennas are optimized for a good impedance match in an infinite planar array environment and are then deformed to fit along the surface of a hemisphere. In summary, circularly polarized or polarization reconfigurable conformal array antennas are optimal choices for mobile platform antennas.

1.2.3.3 Direction of Arrival (DOA) Estimation and Beamforming Algorithms Design

In harsh emergency communication environments, there exist problems such as severe occlusion, high path loss and complex multipath. Air-ground communication based on flight carriers such as balloons and unmanned aerial vehicles can become an important means to solve the above problems. In order to explore the huge potential of air-ground communication in harsh environments, the research on conformal array signal processing technology including DOA estimation algorithm [69–81], sparse optimization algorithm [82–91] and interference suppression beamforming algorithm [92–100] has become an important way to improve the performance of wireless communication.

In a multipath communication environment, coherent signal components are generated when signals propagate through multipaths. These complex coherent signals pose significant challenges to DOA estimation. The traditional DOA estimation algorithm is mainly based on the principle of orthogonal signal subspaces. For example, multiple signal classification (MUSIC) [101] and the estimation of signal parameters via rotational invariance techniques (ESPRIT) [102], these algorithms are mainly aimed at the study of incoherent signals. It is unable to effectively estimate the coherent signal situation. In recent years, thanks to the development of optimization theory, DOA estimation algorithms based on the idea of sparse reconstruction have begun to emerge, such as the L1-SVD convex optimization algorithm [103] and the matching pursuit algorithm (MP) [104]. Different from subspace type algorithms, the DOA estimation algorithm based on the idea of sparse reconstruction transforms the DOA estimation problem into an optimization analysis problem independent of the coherence of the signal, and can effectively estimate the DOA of coherent signals. However, the DOA estimation algorithm based on the idea of sparse reconstruction is generally only applicable to the design of uniform linear arrays and is difficult to be used in the case of conformal arrays such as circular arrays. Studying the DOA estimation algorithm of coherent signals based on conformal arrays has important academic significance and application value. At present, there are relatively few studies on two-dimensional DOA estimation algorithms for coherent signals based on circular arrays. Moreover, the computational complexity required for two-dimensional coherent signal DOA estimation is about two orders of magnitude higher than that of one-dimensional estimation, and decoherence operations are also needed simultaneously. This poses a serious obstacle to the efficient two-dimensional DOA estimation of coherent signals.

Conformal arrays have the advantages of good aerodynamics, saving carrier space, expanding the array deployment range and beam scanning range, etc. However, the number of array elements in conformal arrays will increase with the increase of array size, resulting in a sharp increase in the load, cost and complexity of the entire antenna system. In addition, the side lobe level of the pattern of some conformal antenna arrays is very high, which will cause the influence of signal power dispersion and increased side lobe interference. In response to the above problems, currently, the method of sparse array elements is mainly adopted to optimize the conformal array layout, reduce the number of array elements in the conformal array, and lower the side lobe level, as well as the cost and complexity of the antenna system [105]. The sparse optimization of conformal arrays requires simultaneously reducing the number of array elements, keeping the main beam direction and gain of the expected signal direction unchanged, and suppressing the side lobe level and other requirements as much as possible. These requirements bring considerable challenges to the sparse optimization work of conformal antenna arrays. Classic sparse optimization algorithms, such as the minimum redundancy matrix method, can achieve a relatively high sparsity rate. However, the antenna array after sparsity often has a relatively high side lobe level. The sparse algorithms based on group optimization algorithm [106–109] can effectively solve the sparse optimization problem of conformal arrays on a global scale. However, these algorithms often have problems such as uneven initial population distribution, large computational complexity, poor ability to escape the local optimal solution, and slow convergence speed in the later stage of the loop. There are few related studies on these problems and they are difficult to solve. It is a huge challenge for the optimization algorithm.

Aiming at the problem of severe multi-user concurrent transmission interference generated in harsh communication environments, which makes it difficult to increase communication capacity, it is urgent to design the interference rejection adaptive beamforming (IRAB) algorithm to suppress multi-user interference. The basic principle of IRAB is to generate spatially oriented beams to directionally suppress interference signals. By aligning the main beam of the pattern generated by the array antenna with the desired signal direction and the zero trap with the direction of the interference signal, the purpose of suppressing the power of the interference signal while maintaining the desired signal reception power unchanged can be achieved, improving communication quality and realizing the effect of space division multiplexing. At present, IRAB algorithms are generally divided into two categories: one is statistical optimal beamforming, and the other is robust beamforming. Statistical optimal beamforming generally requires obtaining the ideal second-order statistical characteristics of the received signal to construct the ideal covariance matrix of the received signal. However, in practical applications, the ideal second-order statistical characteristics of the received signal are unknown and difficult to obtain [110]. Therefore, this paper mainly studies robust beamforming. Robust beamforming is calculated using the estimated values of actual data. However, there are errors between the data estimated values and the theoretical values, and the performance of robust beamforming is generally lower than that of statistically optimal beamforming. To improve the performance of robust beamforming, some scholars have proposed the diagonal loading (DL) algorithm. However, the diagonal loading factor of the DL algorithm can affect the effect of beamforming interference suppression. Improper selection can cause the failure of the algorithm. How to determine the diagonal loading factor based on different signal to interference plus noise ratio (SINR), signal angles and interference angles remains a key issue that puzzles scholars. It is of great significance to study the determination methods of the optimal diagonal loading factor in different situations. Furthermore, the current existing research on DL algorithm mainly focuses on linear arrays. There is insufficient research on the algorithms of conformal sparse arrays, which fails to give full play to the advantages of conformal sparse arrays and the interference suppression ability of beamforming.

In conclusion, the research on DOA estimation and beamforming technology based on conformal sparse arrays carried out in this book has significant academic value, and the development and implementation of the corresponding technology have strong application significance.

References

1. C. Ding et al., "Comparative studies on multi-carrier transmission schemes in mountainous and dense forest environment," Sci. Rep., vol. 12, no. 1, p. 16782, Oct. 2022.
2. Hari, K. V. S., Baum, D. S., Rustako, A. J., Roman, R. S. & Trinkwon, D. "Channel models for fixed wireless applications". *IEEE 802.16 Broadband Wireless Access Working Group*, 2003.

3. L. Yuxin, "An Emergency Communication Vehicle Antenna System Based on Static Communication Antenna and Base Station Antenna," *2020 5th International Conference on Electromechanical Control Technology and Transportation (ICECTT)*, Nanchang, China, 2020, pp. 207–212.
4. C. Zhu and B. Feng, "A Dual-Polarized Antenna Using an Origami Structure for 5G Emergency Communications," *2019 International Workshop on Electromagnetics: Applications and Student Innovation Competition (iWEM)*, Qingdao, China, 2019, pp. 1–2.
5. R. Joshi and P. J. Soh, "Dual-Band Folded-Shorted Patch Antenna for Military Search and Rescue Operations and Emergency Communications," *2018 18th International Symposium on Antenna Technology and Applied Electromagnetics (ANTEM)*, Waterloo, ON, Canada, 2018, pp. 1–2.
6. Guan D F, Ding C, Qian Z P, et al. A siw-based large-scale corporate-feed array antenna[J]. *IEEE Trans. Antennas Propag.*, vol. 63, no. 7, pp. 2969-2976. 2015.
7. Yong W Y, Haddadi A, Emanuelsson T, et al. A bandwidth-enhanced cavity-backed slot array antenna for mmwave fixed-beam applications[J]. *IEEE Antennas Wireless Propag. Lett.*, vol. 19, no. 11, pp. 1924-1928. 2020.
8. Liu J, Vosoogh A, Zaman A U, et al. Design and fabrication of a high-gain 60-ghz cavity-backed slot antenna array fed by inverted microstrip gap waveguide[J]. *IEEE Transactions on Antennas and Propagation.*, vol. 65, no. 4, pp. 2117-2122. 2017.
9. Li Y, Luk K M, "60-ghz substrate integrated waveguide fed cavity-backed aperture-coupled microstrip patch antenna arrays. *IEEE Transactions on Antennas and Propagation.*, vol. 63, no. 3, pp. 1075-1085, Jan. 2015.
10. Yang Q, Gao S, Luo Q, et al. "Cavity-backed slot-coupled patch antenna array with dual slant polarization for millimeter-wave base station applications,". *IEEE Transactions on Antennas and Propagation*, vol. 69, no. 3, pp. 1404-1413, Aug. 2021
11. Awida M H, Suleiman S H, Fathy A E, "Substrate-integrated cavity-backed patch arrays: A low-cost approach for bandwidth enhancement,". *IEEE Trans. Antennas Propag.*, vol. 59, no. 4, pp. 1155–1163, Jau. 2011.
12. Cao B, Wang H, Huang Y, et al. "High-gain L-probe excited substrate integrated cavity antenna array with LTCC-based gap waveguide feeding network for w-band application," *IEEE Trans. Antennas Propag.*, vol. 63, no. 12, pp. 5465-5474,. 2015.
13. Shad S, Mehrpouyan H, "60 ghz waveguide-fed cavity array antenna by multistepped slot aperture," *IEEE Antennas Wireless Propag. Lett.*, vol. 19, pp. 438-442, Jan. 2020.
14. Zhou S, Huang G, Peng Z, et al. "Design of a Wideband Dual-Polarization Full-Corporate Waveguide Feed Antenna Array," *IEEE Trans. Antennas Propag.*, vol. 64, no. 5 pp. 1974-1978, Sep. 2016.
15. Huang G L, Zhou S G, Chio T H, et al. "A low profile and low sidelobe wideband slot antenna array feb by an amplitude-tapering waveguide feed-network," *IEEE Trans. Antennas Propag.*, vol. 63, no.1, pp. 419-423, Oct. 2015.
16. Xu J, Chen Z N, Qing X. "CPW center-fed single-layer siw slot antenna array for automotive radars," *IEEE Trans. Antennas Propag.*, vol. 62, no.9, pp. 4528-4536, Jun. 2014.
17. Chen J, Hu T, Zhao Y T, et al. "Realization of high-gain low-sidelobe wide-sector beam using inductive diaphragms loaded slotted ridge waveguide antenna array for air detection applications,". *IEEE Trans. Antennas Propag.*, vol. 70, no.4, pp. 2698-2707, Oct, 2022.
18. Chang L, Zhang Z, Li Y, et al. "Air substrate slot array based on channelized coplanar waveguide," *IEEE Antennas Wireless Propag. Lett.*, vol. 16: pp. 892-895, Sep. 2017.
19. Xu L H, Guo Y Q, Hu P F, et al. "Bandwidth enhancement of the millimeter-wave microstrip lineararray with loading of shorting pins," *IEEE Trans. Antennas Propag.*, vol. 71, no. 1, pp. 1105-1110, Oct. 2022.
20. Xue Q, Guo J, Liao S, et al. "Planar aperture antenna with high gain and high aperture efficiency for 60-ghz applications," *IEEE Trans. Antennas Propag.*, vol. 65, no. 12, pp. 6262-6273, Jul. 2017.
21. Y. J. Cheng, H. Xu, D. Ma, J. Wu, L. Wang, and Y. Fan, "Millimeter- wave shaped-beam substrate integrated conformal array antenna," *IEEE Trans. Antennas Propag.*, vol. 61, no. 9, pp. 4558–4566, Sep. 2013.

22. M. Mozaffari, W. Saad, M. Bennis, Y.-H. Nam, and M. Debbah, "A tutorial on UAVs for wireless networks: applications, challenges, and open problems," *IEEE Commun. Surveys Tuts.*, vol. 21, no. 3, pp. 2334–2360, Mar. 2019.
23. Y. Zeng, R. Zhang, and T. J. Lim, "Wireless communications with unmanned aerial vehicles: opportunities and challenges," *IEEE Commun. Mag.*, vol. 54, no. 5, pp. 36–42, May 2016.
24. S. H. Alsamhi, F. A. Almalki, F. Afghah, A. Hawbani, A. V. Shvetsov, B. Lee, and H. Song, "Drones' edge intelligence over smart environments in B5G: blockchain and federated learning synergy," *IEEE Trans. Green Commun. Netw.*, vol. 6, no. 1, pp. 295–312, Mar. 2022.
25. S. H. Alsamhi, M. S. Ansari, O. Ma, F. Almalki, and S. K. Gupta, "Tethered balloon technology in design solutions for rescue and relief team emergency communication services," *Disaster Med. Publ. Health Preparedness*, vol. 13, no. 2, pp. 203–210, May 2018.
26. S. H. Alsamhi, F. A. Almalki, O. Ma, M. S. Ansari, and M. C. Angelides, "Performance optimization of tethered balloon technology for public safety and emergency communications," *Telecommun. Syst.*, vol. 75, no. 2, pp. 235–244, May 2019.
27. S. H. Alsamhi, M. S. Ansari, and N. S. Rajput, "Disaster coverage predication for the emerging tethered balloon technology: capability for preparedness, detection, mitigation, and response," *Disaster Med. Publ. Health Preparedness*, vol. 12, no. 2, pp. 222–231, Apr. 2018.
28. S. A. Khaleefa, S. H. Alsamhi, and N. S. Rajput, "Tethered balloon technology for telecommunication, coverage and path loss," *in Proc. IEEE Students Conf. Electr. Electron. Comput. Sci.*, 2014, pp. 1–4.
29. C. A. Balanis, *Antenna Theory: Analysis and Design*, 3rd ed. Hoboken, NJ, USA: Wiley, 2005.
30. Z. N. Chen, D. Liu, H. Nakano, X. Qing, and T. Zwick, *Handbook of Antenna Technologies*. Singapore: Springer, 2016.
31. D. Sun, W. Dou, and L. You, "Application of novel cavity-backed proximity-coupled microstrip patch antenna to design broadband conformal phased array," *IEEE Antennas Wireless Propag. Lett.*, vol. 9, pp. 1010–1013, Oct. 2010.
32. S. Xiao, S. Yang, H. Zhang, Q. Xiao, Y. Chen, and S. W. Qu, "Practical implementation of wideband and wide-scanning cylindrically conformal phased array," *IEEE Trans. Antennas Propag.*, vol. 67, no. 8, pp. 5729–5733, Aug. 2019.
33. Q. Chen, Z. Hu, Z. Shen, and W. Wu, "2–18 GHz conformal low-profile log-periodic array on a cylindrical conductor," *IEEE Trans. Antennas Propag.*, vol. 66, no. 2, pp. 729–736, Feb. 2018.
34. P. Qin, L. Song, and Y. J. Guo, "Beam steering conformal transmitarray employing ultra-thin triple-layer slot elements," *IEEE Trans. Antennas Propag.*, vol. 67, no. 8, pp. 5390–5398, Aug. 2019.
35. Y. Xia, B. Muneer and Q. Zhu, "Design of a full solid angle scanning cylindrical-and-conical phased array antennas," *IEEE Trans. Antennas Propag.*, vol. 65, no. 9, pp. 4645–4655, Sep. 2017.
36. Z. Chen, Z. Shen, J. Zhang, and G. Zhang, "Conformal platform-embedded horn antenna with ultra-wide bandwidth and low profile," *IET Microw., Antennas Propag.*, vol. 13, no. 10, pp. 1711–1718, Apr. 2019.
37. V. Jaeck, L. Bernard, K. Mahdjoubi, K., Sauleau, R., Collardey, S., Pouliguen, P., and Potier. "A switched-beam conformal array with a 3-D beam forming capability in C-band," *IEEE Trans. Antennas Propag.*, vol. 65, no. 6, pp. 2950–2957, Jun. 2017.
38. C.H. Ahn, Y.J. Ren, and K. Chang, "A dual-polarized cylindrical conformal array antenna suitable for unmanned aerial vehicles," *Int. J. RF Microw. Comput. Aided Eng.*, vol. 21, no. 1, pp. 91–98, Nov. 2011.
39. V. Semkin, F. Ferrero, A. Bisognin, J. Ala-Laurinaho, C. Luxey, F. Devillers, and A. V. Raisanen. "Beam switching conformal antenna array for mm-wave communications," *IEEE Antennas Wireless Propag. Lett.*, vol. 15, pp. 28–31, Apr. 2015.
40. B. Zhang, C. Jin, K. Cao, Q. Lv, and R. Mittra, "Cognitive conformal antenna array exploiting deep reinforcement learning method," *IEEE Trans. Antennas Propag.*, vol. 70, no. 7, pp. 5094–5104, Jul. 2022.

41. G. Sanford. "Conformal microstrip phased array for aircraft test with ATS-6," *IEEE Trans. Antennas Propag.*, vol. 26, no. 5, pp 642–646, Sep. 1978.
42. X. Gao, Z. Xia, S. Wang, J. Du and Y. Sui, "Low-cost wind-insensitive ultrawideband balloon antenna for seismic telemetry systems," *IEEE Antennas Wireless Propag. Lett.*, vol. 16, pp. 2832–2835, Sep. 2017.
43. G. M. Chi, B. H. Li, and Y. J. Liu, "Numerical analysis of a novel balloon antenna," in *Proc. Int. Symp. IEEE Antennas Propag. Soc.*, Jun. 2004, vol. 3, pp. 2408–2411.
44. Y. J. Cheng, H. Xu, D. Ma, J. Wu, L. Wang, and Y. Fan, "Millimeter-wave shaped-beam substrate integrated conformal array antenna," *IEEE Trans. Antennas Propag.*, vol. 61, no. 9, pp. 4558–4566, Sep. 2013.
45. H. Yang et al., "Design equations for cylindrically conformal arrays of longitudinal slots," *IEEE Trans. Antennas Propag.*, vol. 64, no. 1, pp. 80–88, Jan. 2016.
46. K. Wincza, S. Gruszczynski, and K. Sachse, "Conformal four-beam antenna arrays with reduced sidelobes," *Electron. Lett.*, vol. 44, no. 3, pp. 174–175, Jan. 2008.
47. Y. M. Cheng, P. Chen, W. Hong, T. Djerafi, and K. Wu, "Substrateintegrated-waveguide beam-forming networks and multibeam antenna arrays for low-cost satellite and mobile systems," *IEEE Antennas Propag Mag.*, vol. 53, no. 6, pp. 18–30, Dec. 2011.
48. A. J. Lockyer, K. H. Alt, J. N. Kudva, R. W. Kinslow, and A. C. Goetz,"Conformal load-bearing antenna structures (CLAS): Initiative for multiple military and commercial applications," *Proc. SPIE*, vol. 3046, pp. 182–196, Jun. 1997.
49. P. Wang, G. Wen, H. Zhang, and Y. Sun, "A wideband conformal end- fire antenna array mounted on a large conducting cylinder," *IEEE Trans. Antennas Propag.*, vol. 61, no. 9, pp. 4857–4861, Sep. 2013.
50. J.-J. Peng, S.-W. Qu, M. Xia, and S. Yang, "Conformal phased array antenna for unmanned aerial vehicle with $\pm 70\circ$ scanning range," *IEEE Trans. Antennas Propag.*, vol. 69, no. 8, pp. 4580–4587, Aug. 2021.
51. A. Y. Svezgentsev, V. Volski, S. Yan, P. J. Soh, and G. A. E. Vandenbosch, "Omnidirectional wideband E-shaped cylindrical patch antennas," *IEEE Trans. Antennas Propag.*, vol. 64, no. 2, pp. 796–800, Feb. 2016.
52. C. Ding, X. Gao, W. Tang, H. Liu, C. Zhu, X. Bu, and J. An, "A balloon conformal heterogeneous array antenna with ultrawide bandwidth and half-space beam coverage," *IEEE Trans. Antennas Propag.*, vol. 72, no. 9, pp. 7077-7088, Sept. 2024.
53. D. Caratelli, G. Toso, O. V. Stukach, and N. V. Panokin, "Deterministic constrained synthesis technique for conformal aperiodic linear antenna arrays—Part II: Applications," *IEEE Trans. Antennas Propag.*, vol. 67, no. 9, pp. 5962–5973, Sep. 2019.
54. M.-C. Tang, Y. Duan, Z. Wu, and M. Li, "Flexible, conformal, endfire radiating, electrically small antennas on a conducting cylinder with a small radius," *IEEE Trans. Antennas Propag.*, vol. 68, no. 5, pp. 3496–3503, May 2020.
55. R. Xu and Z. N. Chen, "A hemispherical wide-angle beamsteering near-surface focal-plane metamaterial Luneburg lens antenna using transformation-optics," *IEEE Trans. Antennas Propag.*, vol. 70, no. 6, pp. 4224–4233, Jun. 2022.
56. Y. Wang and S. -W. Qu, "Dual-polarized conformal transparent antenna array with hemispherical beam coverage," *IEEE Antennas Wireless Propag. Lett.*, vol. 22, no. 10, pp. 2452-2456, Oct. 2023.
57. H. Zhu, X. Liang, S. Ye, R. Jin, and J. Geng, "A cylindrically conformal array with enhanced axial radiation," *IEEE Antennas Wireless Propag. Lett.*, vol. 15, pp. 1653–1656, 2016.
58. A. T. Castro and S. K. Sharma, "Inkjet-printed wideband circularly polarized microstrip patch array antenna on a PET film flexible substrate material," *IEEE Antennas Wireless Propag. Lett.*, vol. 17, no. 1, pp. 176–179, Jan. 2018.
59. G. Chenhu, J. Geng, H. Zhou, J. Li, L. Liu, and X. Liang, "Truncated circular cone slot antenna array that radiates a circularly polarized conical beam," *IEEE Antennas Wireless Propag. Lett.*, vol. 16, pp. 2574–2577, 2017.
60. W. Tang, X. Gao, C. Ding, C. Zhu, R. Li, B. Tian, X. Bu, and J. An, "A quad-polarization reconfigurable conformal array of wide coverage range and high gain for unmanned aerial

vehicle communications," *IEEE Trans. Antennas Propag.*, vol. 72, no. 4, pp. 3795-3800, April 2024.

61. S. Ogurtsov and S. Koziel, "A conformal circularly polarized series-fed microstrip antenna array design," *IEEE Trans. Antennas Propag.*, vol. 68, no. 2, pp. 873–881, Feb. 2020.
62. H. S. Lin, Y. J. Cheng, Y. F. Wu, and Y. Fan, "Height reduced concave sector-cut spherical conformal phased array antenna based on distributed aperture synthesis," *IEEE Trans. Antennas Propag.*, vol. 69, no. 10, pp. 6509–6517, Oct. 2021.
63. C. Yu, S. Yang, Y. Han, L. Jin, Y. Chen, W. Wang, Z. Li, L. Zhang, and Y. Zheng, "A DBDCP antenna with a helmet-conformal AMC for industrial IoT applications featuring LHCP and RHCP in the low and high bands, respectively," *IEEE Internet of Things Journal.*, vol. 11, no. 12, pp. 22310-22320, Jun, 2024.
64. Y. Fan, X. Quan, Y. Pan, Y. Cui, and R. Li, "Wideband omnidirectional circularly polarized antenna based on tilted dipoles," *IEEE Trans. Antennas Propag.*, vol. 63, no. 12, pp. 5961–5966, Dec. 2015.
65. C. Yang, X. Y. Liu, and Y. Fan, "Design of broadband circularly polarized all-textile antenna and its conformal array for wearable devices," *IEEE Trans. Antennas Propag.*, vol. 70, no. 1, pp. 209–220, Jan. 2022.
66. Y.-H. Huang, Y. Wang, Y.-L. Zou, and J.-L. Guo, "A miniature circularly polarized air-borne antenna with wide angle coverage," *IEEE Antennas Wireless Propag. Lett.*, vol. 16, pp. 497–500, 2017.
67. Gao, W. Jiang, W. Hu, Q. Wang, W. Zhang, and S. Gong, "A dual-polarized two-dimensional monopulse antenna array for conical conformal applications," *IEEE Trans. Antennas Propag.*, vol. 69, no. 9, pp. 5479–5488, Sep. 2021.
68. C. Pfeiffer and J. Massman, "A UWB low-profile hemispherical array for wide angle scanning," *IEEE Trans. Antennas Propag.*, vol. 71, no. 1, pp. 508–517, Jan. 2023.
69. J. Capon, "High-resolution frequency-wavenumber spectrum analysis," in *Proceedings of the IEEE*, vol. 57, no. 8, pp. 1408-1418, Aug. 1969.
70. M. D. Zoltowski, "On the performance analysis of the MVDR beamformer in the presence of correlated interference," in *IEEE Transactions on Acoustics, Speech, and Signal Processing*, vol. 36, no. 6, pp. 945-947, June 1988.
71. M. Zhang, J. -y. Zhang, J. -l. Jin, G. -q. Wang and R. -m. He, "Maximum entropy spectral estimation based on accelerating genetic algorithm," *2009 Chinese Control and Decision Conference*, Guilin, China, 2009, pp. 4184–4189.
72. J. Zhou, L. Lei and N. Liu, "Research on rough and detailed combination spectrum analysis of electromagnetic signals based on complex modulation ZoomFFT," *2019 IEEE 19th International Conference on Communication Technology (ICCT)*, Xi'an, China, 2019, pp. 786–789.
73. M. Rubsamen and A. B. Gershman, "Performance analysis of root-music-based direction-of-arrival estimation for arbitrary non-uniform array," *2008 5th IEEE Sensor Array and Multichannel Signal Processing Workshop*, Darmstadt, Germany, 2008, pp. 381–385.
74. P. Tripathy, S. C. Srivastava and S. N. Singh, "A modified TLS-ESPRIT-Based method for low-frequency mode identification in power systems utilizing synchrophasor measurements," in *IEEE Transactions on Power Systems*, vol. 26, no. 2, pp. 719-727, May 2011.
75. C. Chen and X. Zhang, "A RD-ESPRIT algorithm for coherent DOA estimation in monostatic MIMO radar using a single pulse," in *International Journal of Electronics*, vol. 101, no. 8, pp, 1074–1085. 2013.
76. L. Wang, K. Lu and P. Liu, "Compressed sensing of a remote sensing image based on the priors of the reference image," in *IEEE Geoscience and Remote Sensing Letters*, vol. 12, no. 4, pp. 736-740, April 2015.
77. J. Yin and T. Chen, "Direction-of-Arrival estimation using a sparse representation of array covariance vectors," in *IEEE Transactions on Signal Processing*, vol. 59, no. 9, pp. 4489-4493, Sept. 2011.
78. Tie-Jun Shan and T. Kailath, "Adaptive beamforming for coherent signals and interference," in *IEEE Transactions on Acoustics, Speech, and Signal Processing*, vol. 33, no. 3, pp. 527–536, June 1985.

79. Tie-Jun Shan, M. Wax and T. Kailath, “On spatial smoothing for direction-of-arrival estimation of coherent signals,” in *IEEE Transactions on Acoustics, Speech, and Signal Processing*, vol. 33, no. 4, pp. 806–811, August 1985.
80. R. T. Williams, S. Prasad, A. K. Mahalanabis and L. H. Sibul, “An improved spatial smoothing technique for bearing estimation in a multipath environment,” in *IEEE Transactions on Acoustics, Speech, and Signal Processing*, vol. 36, no. 4, pp. 425-432, April 1988.
81. R. Eiges and H. Griffiths, “Mode-space spatial spectral estimation for circular arrays,” in *IEE Proceedings-Radar, Sonar and Navigation*, vol. 141, no. 6, pp. 300-306, 1994.
82. Z. -C. Hao, M. He and W. Hong, “Design of a millimeter-wave high angle selectivity shaped-beam conformal array antenna using hybrid genetic/space mapping method,” in *IEEE Antennas and Wireless Propagation Letters*, vol. 15, pp. 1208-1212, 2016.
83. H. Chireix, “High power outphasing modulation,” in *Proceedings of the Institute of Radio Engineers*, vol. 23, no. 11, pp. 1370-1392, Nov. 1935.
84. M. Kanno, T. Hashimura, T. Katada, M. Sato, K. Fukutani and A. Suzuki, “Digital beam forming for conformal active array antenna,” *Proceedings of International Symposium on Phased Array Systems and Technology*, Boston, MA, USA, 1996, pp. 37–40.
85. P. Knott, “Antenna design and beamforming for a conformal antenna array demonstrator,” *2006 IEEE Aerospace Conference*, Big Sky, MT, USA, 2006, pp. 7 pp.-
86. Y. -Y. Bai, S. Xiao, C. Liu and B. -Z. Wang, “A hybrid IWO/PSO algorithm for pattern synthesis of conformal phased arrays,” in *IEEE Transactions on Antennas and Propagation*, vol. 61, no. 4, pp. 2328-2332, April 2013.
87. Q. Ling, C. Yang, Y. Yu, T. Liu and X. Sheng, “Sparsity and adaptivity for sensor location optimization of underwater conformal vector array,” *Global Oceans 2020: Singapore – U.S. Gulf Coast*, Biloxi, MS, USA, 2020, pp. 1–5.
88. Liang, L., Jin, C., Li, H., Liu, J., Jiang, Y. and Zhou, J., “A hybrid algorithm of orthogonal perturbation method and convex optimization for beamforming of sparse antenna array,” in *Electromagnetics*, vol. 4, no. 40, pp. 227–243, 2020.
89. R. Harrington, “Sidelobe reduction by nonuniform element spacing,” in *IRE Transactions on Antennas and Propagation*, vol. 9, no. 2, pp. 187-192, March 1961.
90. M. Skolnik, J. Sherman and F. Ogg, “Statistically designed density-tapered arrays,” in *IEEE Transactions on Antennas and Propagation*, vol. 12, no. 4, pp. 408-417, July 1964.
91. Y. Wang, Y. Zhao and X. Zhang, “Adaptive phase switching snake optimization based on orthogonal learning and joint opposite selection,” *2023 IEEE 5th International Conference on Power, Intelligent Computing and Systems (ICPICS)*, Shenyang, China, 2023, pp. 316–322.
92. B. Widrow, P. E. Mantey, L. J. Griffiths and B. B. Goode, “Adaptive antenna systems,” in *Proceedings of the IEEE*, vol. 55, no. 12, pp. 2143-2159, Dec. 1967.
93. A. Pezeshki, B. D. Van Veen, L. L. Scharf, H. Cox and M. Lundberg Nordenvaad, “Eigenvalue beamforming using a multirank MVDR beamformer and subspace selection,” in *IEEE Transactions on Signal Processing*, vol. 56, no. 5, pp. 1954–1967, May 2008.
94. I. S. Reed, J. D. Mallett and L. E. Brennan, “Rapid convergence rate in adaptive arrays,” in *IEEE Transactions on Aerospace and Electronic Systems*, vol. AES-10, no. 6, pp. 853–863, Nov. 1974.
95. R. G. Lorenz and S. P. Boyd, “Robust minimum variance beamforming,” in *IEEE Transactions on Signal Processing*, vol. 53, no. 5, pp. 1684-1696, May 2005.
96. J. P. Lie, W. Ser and C. M. S. See, “Adaptive uncertainty based iterative robust capon beamformer using steering vector mismatch estimation,” in *IEEE Transactions on Signal Processing*, vol. 59, no. 9, pp. 4483-4488, Sept. 2011.
97. Sturm J F., “Using SeDuMi 1.02, A MATLAB toolbox for optimization over symmetric cones,” in *Optimization methods and software*, vol. 1-4, no. 11, pp. 625-653, 1999.
98. Lobo M S, Vandenberghe L, Boyd S, et al. “Applications of second-order cone programming,” in *Linear algebra and its applications*, vol. 1-3, no. 284, pp. 193-228, 1998.
99. K. Senne, “Adaptive array principles,” in *IEEE Antennas and Propagation Society Newsletter*, vol. 25, no. 4, pp. 25-26, August 1983.

100. S. Shahbazpanahi, A. B. Gershman, Zhi-Quan Luo and Kon Max Wong, “Robust adaptive beamforming for general-rank signal models,” in *IEEE Transactions on Signal Processing*, vol. 51, no. 9, pp. 2257–2269, Sept. 2003.
101. R. Schmidt, “Multiple emitter location and signal parameter estimation,” in *IEEE Transactions on Antennas and Propagation*, vol. 34, no. 3, pp. 276-280, March 1986.
102. A. Paulraj, R. Roy and T. Kailath, “A subspace rotation approach to signal parameter estimation,” in *Proceedings of the IEEE*, vol. 74, no. 7, pp. 1044-1046, July 1986.
103. D. Malioutov, M. Cetin and A. S. Willsky, “A sparse signal reconstruction perspective for source localization with sensor arrays,” in *IEEE Transactions on Signal Processing*, vol. 53, no. 8, pp. 3010-3022, Aug. 2005.
104. S. G. Mallat and Zhifeng Zhang, “Matching pursuits with time-frequency dictionaries,” in *IEEE Transactions on Signal Processing*, vol. 41, no. 12, pp. 3397–3415, Dec. 1993.
105. Z. Y. Bao and J. Z. Yu, “Thinning curved surface arrays: Application of the genetic algorithm (GA),” *IET International Radar Conference 2015*, Hangzhou, 2015, pp. 1-4.
106. M N Fakher, “Pattern synthesis for a conformal array antenna mounted on a paraboloid reflector using genetic algorithms,” in *Applied Computational Electromagnetics Society Journal*, vol. 2, no. 30, pp. 230-236, 2015.
107. M. M. S. Taheri, A. R. Mallahzadeh and A. Foudazi, “Shaped beam synthesis for shaped reflector antenna using PSO algorithm,” *2012 6th European Conference on Antennas and Propagation (EUCAP)*, Prague, Czech Republic, 2012, pp. 1–5.
108. J. -L. Guo and J. -Y. Li, “Pattern Synthesis of Conformal Array Antenna in the Presence of Platform Using Differential Evolution Algorithm,” in *IEEE Transactions on Antennas and Propagation*, vol. 57, no. 9, pp. 2615-2621, Sept. 2009.
109. F A Hashim and A G Hussien, “Snake optimization: A novel meta-heuristic optimization algorithm,” in *Knowledge-Based Systems*, vol. 242, pp. 108320, 2022.
110. B. D. Carlson, “Covariance matrix estimation errors and diagonal loading in adaptive arrays,” in *IEEE Transactions on Aerospace and Electronic Systems*, vol. 24, no. 4, pp. 397-401, July 1988.

Chapter 2
Balloon Conformal Array Antennas

Due to their characteristics of low cost, longer hang time and smaller volume after deflation, inflatable balloons are well-suited as aerial base stations for emergency communication applications. Consequently, designing high-performance balloon-conformal array antennas is critical for diverse emergency communication requirements. This chart presents three balloon-conformal array antennas for emergency communication applications, progressively improving beam coverage from directional to half-space beam coverage.

2.1 A Balloon Conformal Log-Periodic Array Antenna

2.1.1 Comparative Analysis of Conformal Strip-Dipole and Bowtie-Dipole

Before we describe the design of conformal log-periodic balloon antenna, it is necessary to investigate and compare the performance of different conformal dipoles first. Figure 2.1 shows the models of a conformal strip-dipole and bowtie-dipole in ANSYS High Frequency Structure Simulator (HFSS), which can be conformal with a cylindrical balloon of diameter D.

The specific configuration parameters of the conformal dipoles shown in Fig. 2.1 are listed in Table 2.1. Here, the curvature radius ($D/2$) of two conformal dipoles are both set as 0.5 m, which is consistent with the design of the conformal log-periodic balloon antenna to be described later. The fundamental parameters of the conformal strip-dipole are chosen as $L = 2.5$ m and $w = 39.3$ mm, which corresponds to the size of a cylindrical dipole operating at 60 MHz and a length-to-diameter ratio of 100. In comparison, the total length of the conformal bowtie-dipole is 2 m, which is only 80% of that of the conformal strip-dipole.

X. Gao et al., *Conformal Array Antennas and Beamforming Technologies for Emergency Communication Applications*, Modern Antenna,
https://doi.org/10.1007/978-981-95-2905-6_2

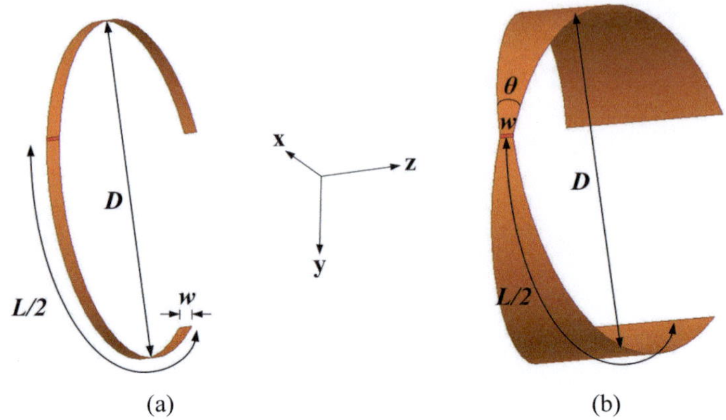

Fig. 2.1 Models of two conformal dipoles in ANSYS HFSS. **a** conformal strip-dipole. **b** Conformal bowtie-dipole

Figure 2.2 shows the simulated input impedance results of the conformal strip-dipole and bowtie-dipole. It can be clearly seen that the conformal bowtie-dipole exhibits the same resonant frequency (near 60 MHz) as the conformal strip-dipole in spite of a 20% reduction in its total length. In addition, compared with the conformal strip-dipole, the conformal bowtie-dipole has flatter curves with regard to the operating frequency, indicating a wider bandwidth.

Figure 2.3 illustrates a comparison of the simulated radiation patterns on E-plane (yz-plane) between these two conformal dipoles. It is observed that the conformal bowtie-dipole has a narrower main-lobe and radiates less power in the y direction compared with its counterpart, which mainly results from the differences of electric current distribution between two dipoles. As shown in Fig. 2.4, the leftmost and rightmost currents on the conformal strip-dipole are nearly equal in amplitude but opposite in direction because of the severe bending, which results in a cancellation in the far-region radiated fields along the z direction. In comparison, although there are also some opposite electric currents on the conformal bowtie-dipole, their values are significantly smaller than those of the conformal strip-dipole, since they are much closer to the open ends due to the size reduction.

Figure 2.5 shows the simulated directivity of the conformal strip-dipole and bowtie-dipole along the z direction. It is found that the directivity of the conformal

Table 2.1 Configuration parameters of conformal dipoles

Parameters	Conformal strip-dipole	Conformal bowtie-dipole
D	1.0 m	1.0 m
L	2.5 m	2.0 m
w	39.3 mm	39.3 mm
θ	–	20°

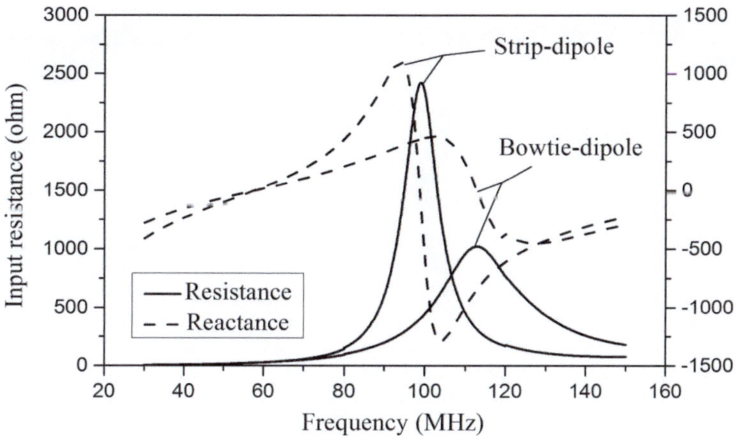

Fig. 2.2 Simulated input impedance of conformal strip-dipole and bowtie-dipole

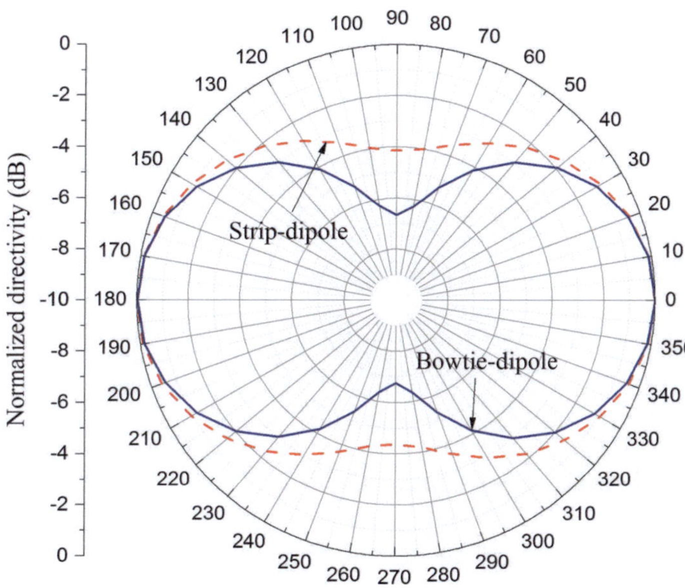

Fig. 2.3 Simulated radiation patterns on *E*-plane (*yz*-plane) of conformal strip-dipole and bowtie-dipole

dipoles decreases as the frequency increases, which is mainly attributed to larger electric currents appearing on the leftmost and rightmost parts of the dipoles at higher frequencies. Obviously, the conformal bowtie-dipole has a higher and less-sensitive directivity compared with the conformal strip-dipole.

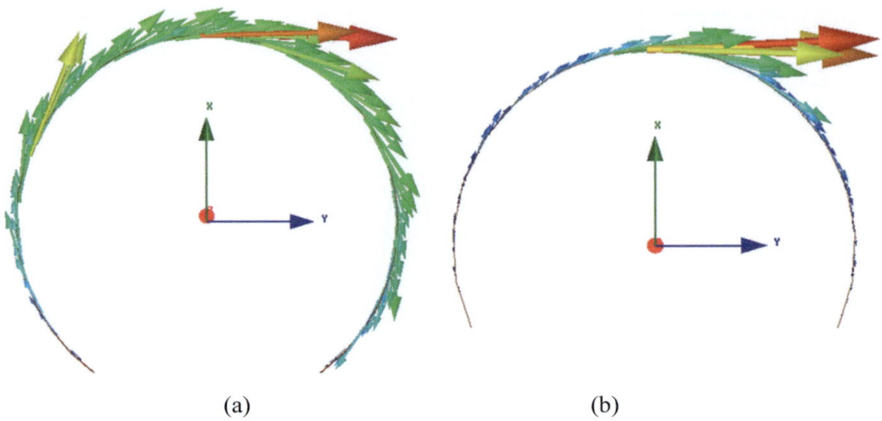

Fig. 2.4 Simulated electric current distribution along conformal dipoles. **a** Conformal strip-dipole. **b** Conformal bowtie-dipole

Fig. 2.5 Simulated directivity of conformal strip-dipole and bowtie-dipole along the *z* direction

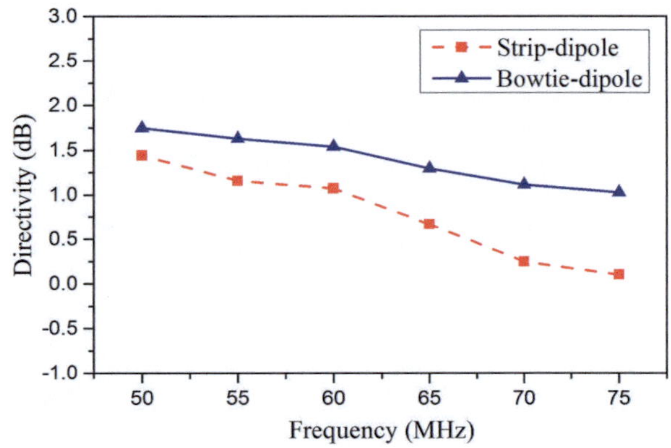

2.1.2 Conformal Log-Periodic Balloon Array Antenna

2.1.2.1 Design of Log-Periodic Balloon Array Antenna

In this section, a conformal log-periodic balloon antenna is designed at VHF band. As shown in Fig. 2.6, the antenna contains eight bowtie-dipole elements, which are conformal with the surface of a cylindrical balloon of 1 m in diameter and 2 m in length. As demonstrated in Section II, conformal bowtie-dipole is employed in the design of log-periodic balloon antenna due to its superior performances. The bowtie-dipoles are cross-fed by a parallel-strip transmission line, printed on the top and bottom layer of a thin Rogers 4003C substrate with dielectric constant of 3.55 and thickness of 1.5 mm. Moreover, a terminal resistor with 50 ohms is connected to the open-end of the balloon antenna to absorb the residual power.

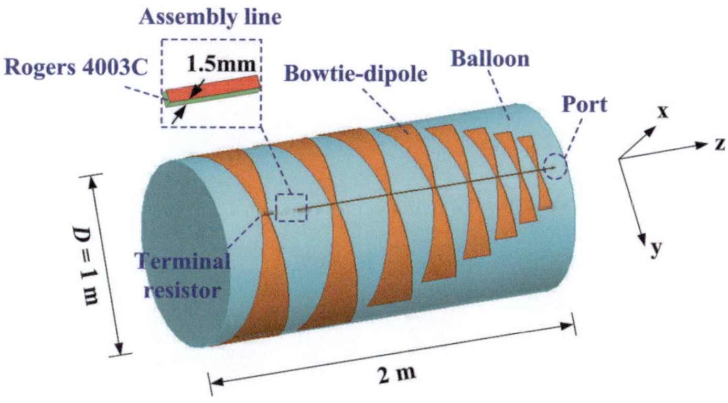

Fig. 2.6 Geometry of a conformal log-periodic balloon antenna

Figure 2.7 shows the expanded planar view of the conformal log-periodic balloon antenna. According to 1, the gain of log-periodic antenna mainly depends upon the scaling factor τ and spacing factor σ. Generally, choosing higher values of τ and σ gives a relatively higher gain, yet results in a bigger antenna size and a greater number of elements. In order to miniaturize the size of the balloon antenna, the scaling factor τ is set as 0.8. For τ smaller than 0.8, the performance of the antenna would deteriorate since there is only about one element in the active region [1]. Moreover, the spacing factor σ is chosen as 0.08, which together with τ brings about a theoretical gain of about 6 dB according to 1. In addition, the flare angle of the conformal bowtie-dipoles is set as 20° due to the limited space. Using the design guidelines in 1 and fine-tune through HFSS simulations, the final antenna parameters are obtained and listed in Table 2.2.

Figure 2.8 shows the simulated and optimized VSWR of the conformal log-periodic balloon antenna. Since the log-periodic antenna can be considered as a transmission line with capacitive loading, its input impedance greatly depends on the characteristic impedance of the central parallel-strip line. As shown in Fig. 2.8, the optimum impedance matching is obtained when the width of the parallel-strip is 3.5 mm. It can be clearly seen that in spite of the relatively small size and few elements, the optimized antenna has a very wide impedance bandwidth ranging from 60 to 233 MHz with VSWR less than two.

Figure 2.9 shows the simulated *E*-plane (*yz*-plane) and *H*-plane (*xz*-plane) radiation patterns of the conformal log-periodic balloon antenna. Due to the size restriction, the longest bowtie-dipole resonates at 60 MHz, which results in a large back-lobe of the end-fire radiation pattern at 60 MHz. It might be reduced if another longer element could be added. As the operating frequency increases, the back-lobe decreases rapidly and the front-to-back ratio achieves 10 dB at 80 MHz.

In order to describe the flexibility of the proposed balloon antenna for practical applications, it is necessary to further investigate the influence of the cylindrical curvature on the antenna's radiation performance. Figure 2.10 shows the simulated

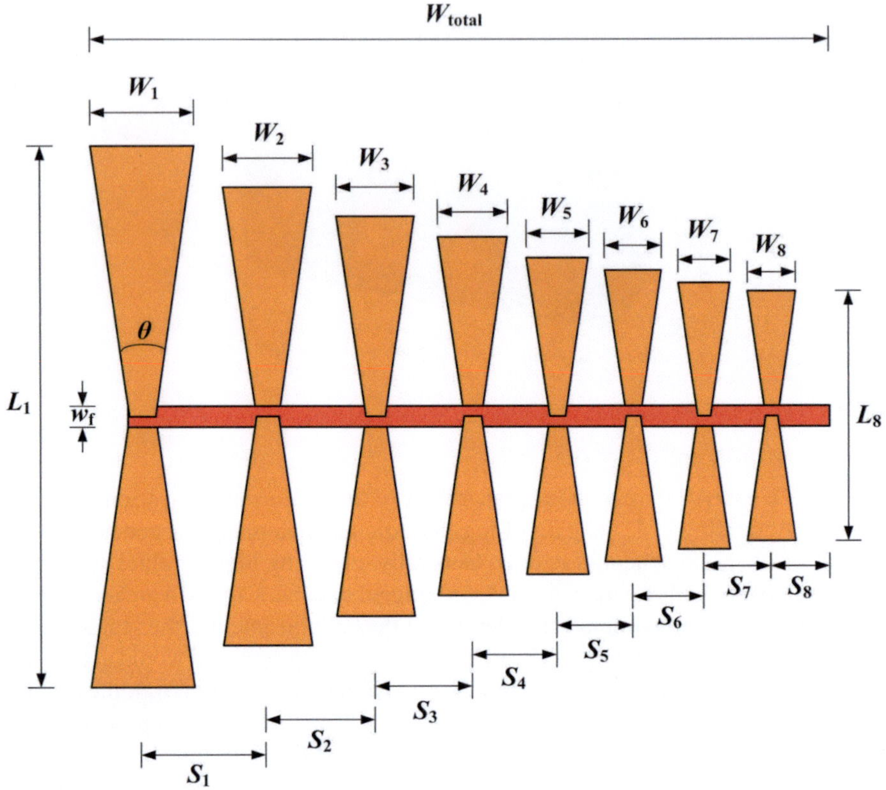

Fig. 2.7 Expanded planar view of the conformal log-periodic balloon antenna

Table 2.2 Geometric parameters of the balloon antenna

Symbol	Value	Symbol	Value	Symbol	Value
L_1	2.000 m	W_1	39.2 cm	S_1	40.0 cm
L_2	1.600 m	W_2	31.4 cm	S_2	32.0 cm
L_3	1.280 m	W_3	25.1 cm	S_3	25.6 cm
L_4	1.024 m	W_4	20.1 cm	S_4	20.5 cm
L_5	0.819 m	W_5	16.1 cm	S_5	16.4 cm
L_6	0.655 m	W_6	12.8 cm	S_6	13.1 cm
L_7	0.524 m	W_7	10.3 cm	S_7	10.5 cm
L_8	0.419 m	W_8	8.2 cm	S_8	8.4 cm
θ	20°	W_{total}	1.860 m	w_f	3.5 mm

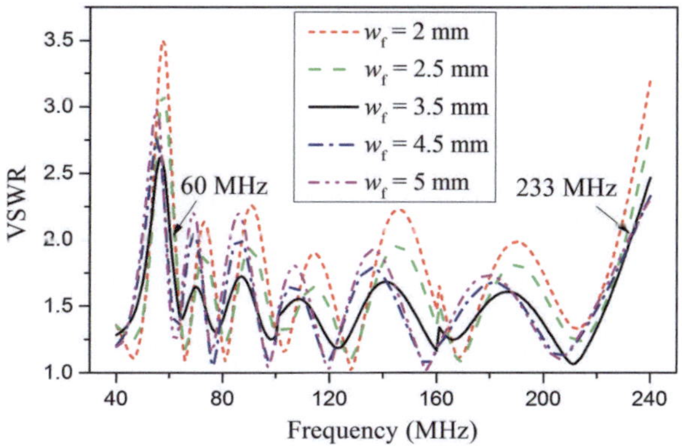

Fig. 2.8 Simulated and optimized VSWR of the conformal log-periodic balloon antenna

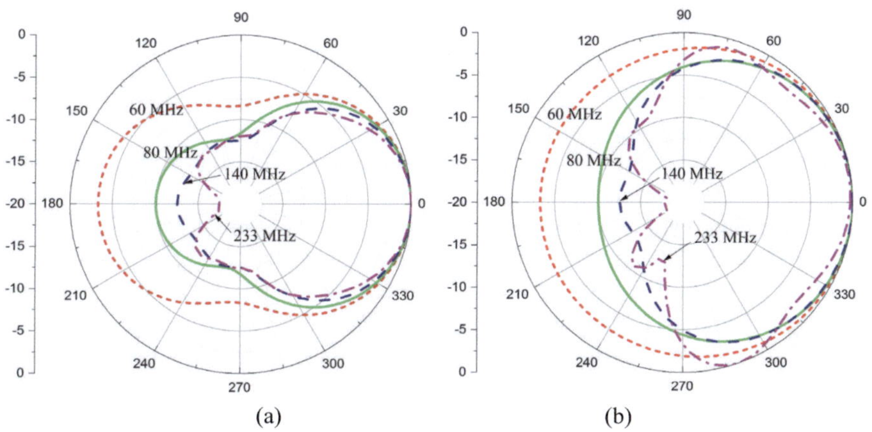

Fig. 2.9 Simulated radiation patterns of the conformal log-periodic balloon antenna. **a** *E*-plane (*yz*-plane). **b** *H*-plane (*xz*-plane)

input impedance of the conformal log-periodic balloon antenna with different diameters. It can be clearly seen that the input resistance and reactance of the antenna fluctuate more gently as the balloon diameter *D* increases. For smaller values of *D*, the curves tend to be more frequency dependent, especially at the low band, which is mainly attributed to the fact that the miniaturized antenna elements exhibit higher quality factor and weaker radiation ability when they are severely bent. Figure 2.11 illustrates the effect of balloon diameter on the radiation patterns of the balloon antenna, which clearly shows that the front-to-back ratio at the low band deteriorates rapidly as *D* decreases in low frequencies. To summarize, the proposed balloon antenna is able to operate with a good radiation performance for balloon diameter more than 1 m.

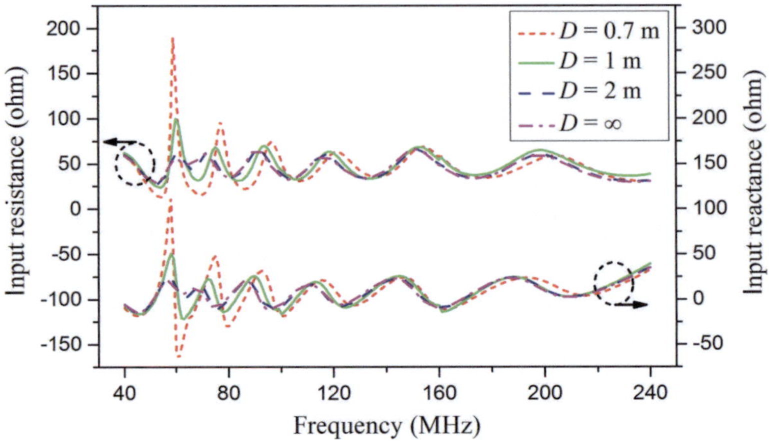

Fig. 2.10 Simulated input impedance of the conformal log-periodic balloon antenna with different diameters D

As a special type of log-periodic dipole antenna, the proposed conformal log-periodic balloon antenna shown in Fig. 2.6 has a number of advantages over traditional structures. First of all, unlike conventional VHF log-periodic antennas commonly made up of long and straight metal cylinders or tubes, our proposed antenna structure can be implemented through attaching pre-cut thin aluminum or copper foils with a PVC balloon, which is very easy to fabricate and cost-effective accordingly. In addition, compared with the traditional VHF log-periodic antennas, the proposed balloon antenna is very light and can be easily packaged into a small volume for transport. Finally, the balloon antenna can be readily deplored to a flexible height from the ground, removing the hassle of establishing a large installation platform. This is particularly suitable for some wireless applications such as temporary or emergency communications on the sea or in a remote area.

2.1.3 Measurement of the Balloon Array Antenna

Figure 2.12 shows the experimental configuration of the balloon antenna for gain measurement. It is seen that two identical conformal log-periodic balloon antennas are used as transmitting and receiving antennas, and they are placed above the earth surface at the same height. Moreover, two radio frequency (RF) baluns with low insertion loss (FTB-1-1 + from Mini-Circuits), are employed to realize the unbalance-to-balance transition at both ports of these balloon antennas. Through measuring the transmission parameter S_{21} with a vector network analyzer (VNA), the gain of the balloon antenna can be obtained accordingly. As for the VSWR measurements, the results can be directly acquired from the VNA with the similar experimental configuration.

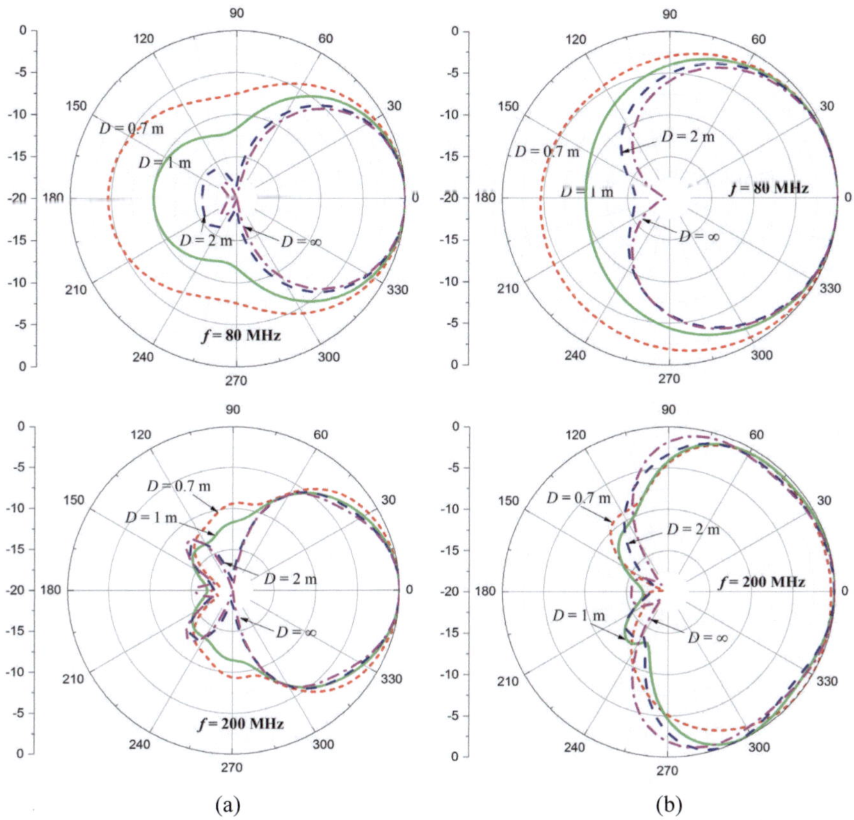

Fig. 2.11 Simulated radiation patterns of the conformal log-periodic balloon antenna with different diameters D at different frequencies. **a** E-plane (yz-plane). **b** H-plane (xz-plane)

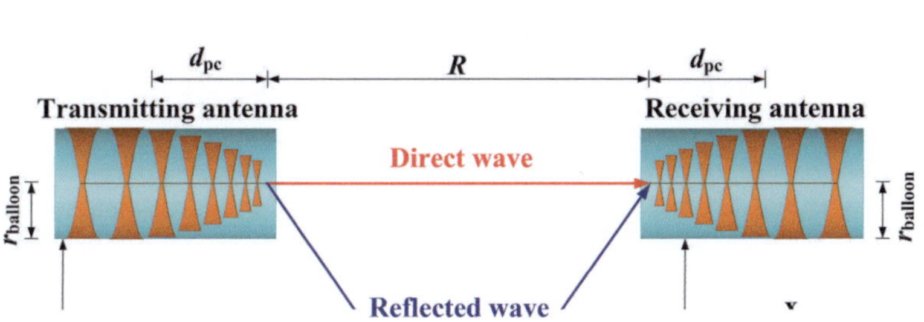

Fig. 2.12 Experimental configuration of the balloon antenna for gain measurement

As shown in Fig. 2.12, since there exists an infinite earth ground below the balloon antenna, it is necessary to investigate the effect of the earth ground on the antenna's radiation performance. Assuming that the earth layer is made of soil with dielectric constant of 15 and conductivity of 0.028 S/m^3, Fig. 2.13 shows the simulated VSWR of the balloon antenna above an infinite earth ground at various heights from 0.5 to 4 m. It can be clearly seen that for h larger than 2 m, the VSWR is very stable and similar to the one obtained in the case of free space. Figure 2.14 shows the radiation efficiency of the conformal log-periodic balloon antenna above an infinite earth ground. To avoid the degradation of radiation efficiency caused by ground loss, the balloon should be placed at least 2 m away from the earth ground. With a sufficient height, the radiation efficiency of the balloon antenna is relatively high and stable, larger than 80% from 60 to 233 MHz.

As for the gain measurement illustrated in Fig. 2.12, it is unrealistic to calculate the scattering parameter S_{21} with HFSS due to the limited memory of our computers. Here, an alternative approach is developed to efficiently model the actual measurement situation. As shown in Fig. 2.12, due to the soil ground, there are two waves propagating toward the receiving balloon antenna. Considering both polarized components, the measured S_{21} can be derived as follows [2].

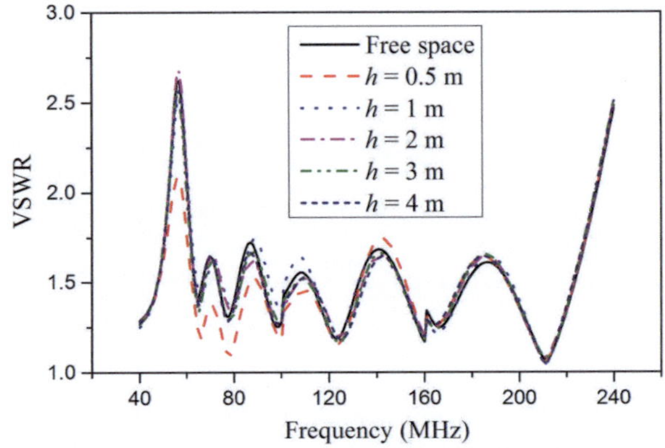

Fig. 2.13 Simulated VSWR of the balloon antenna above an earth

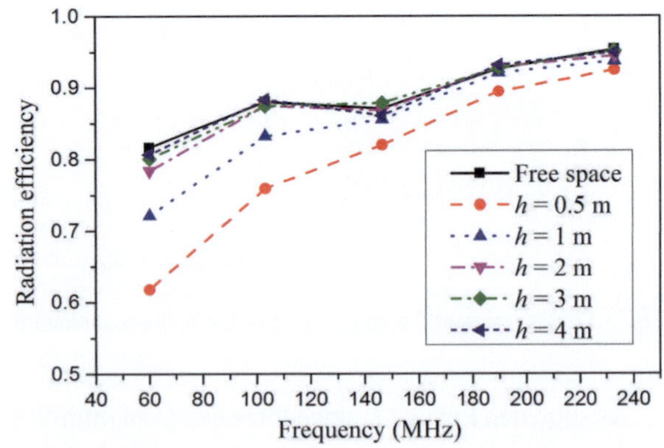

Fig. 2.14 Simulated radiation efficiency of the balloon antenna above an earth

$$S_{21}=\frac{\lambda\exp\left[-jk\left(R+2d_{\rm pc}\right)\right]}{4\pi\left(R+2d_{\rm pc}\right)}\left(G_\theta|_{\theta=0}+G_\varphi|_{\theta=0}\right)$$
$$+\frac{\lambda\exp\left(-jk\sqrt{\left(R+2d_{\rm pc}\right)^2+4(h+r_{\rm balloon})^2}\right)}{4\pi\sqrt{\left(R+2d_{\rm pc}\right)^2+4(h+r_{\rm balloon})^2}}$$
$$\cdot\left(\rho_\theta G_\theta\Big|_{\substack{\theta=\pi/2-\theta_i\\ \varphi=\pi/2}}+\rho_\varphi G_\varphi\Big|_{\substack{\theta=\pi/2-\theta_i\\ \varphi=\pi/2}}\right) \tag{2.1}$$

where R is the distance between the feed ports of two balloon antennas, $d_{\rm pc}$ is the distance between the phase center and the feed port, $(\rho_\theta, \rho_\varphi)$ are the reflection coefficients of the soil ground for the two polarizations, and (G_θ, G_φ) represent the realized gains of the balloon antenna in free space. Here, both $d_{\rm pc}$ and (G_θ, G_φ) can be readily obtained using HFSS. After S_{21} is calculated, the gain of the balloon antenna can then be derived using the following formula [2],

$$G = 10\log_{10}\left[4\pi\left(R+2d_{\rm pc}\right)/\lambda\right]+10\log_{10}(|S_{21}|) \tag{2.2}$$

Based on (2.1) and (2.2), the gain of the balloon antennas with port-to-port distance of 10 m is calculated and shown in Fig. 2.15. The results demonstrate that as the height h is 15 m or higher, the gain is nearly unchanged for all frequencies and is almost identical to the one obtained in free space. The main reason is that the received signal reflected from the earth ground is much smaller than the one from the direct path at larger heights. However, although larger heights result in better results, it is inconvenient to construct such a high experimental configuration in practical environment. In fact, although discrepancy occurs for heights from 2 to 10 m, the error in gain measurement is small and less than 1 dB.

Considering the effect of the earth ground as well as the practical measurement set-up, the height h and port-to-port distance R for conformal log-periodic balloon antennas are chosen as 2 m and 10 m, respectively. Figure 2.16 shows the set-up

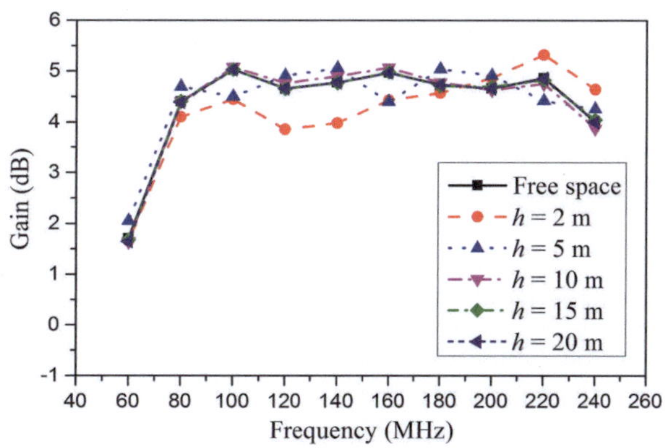

Fig. 2.15 Calculated gain for balloon antennas above the earth ground

photograph of the balloon antennas, which are fabricated by cutting aluminum foil tapes and sticking them to cylindrical balloons made of PVC material. Figures 2.17 and 2.18 show the measured VSWR and realized gain of the balloon antenna, respectively. It can be found that measured results are in good agreement with simulated ones. The operating frequency range of our designed conformal log-periodic balloon antenna is from 60 to 233 MHz, and the average measured gain is around 4.5 dBi. Here, since the balloon antennas have relatively large structural sizes, it is extremely difficult and inconvenient to construct two big nonmetallic rotation platforms for degree-by-degree measurements of radiation patterns. Nevertheless, the measured results of realized gain indicates that the balloon antenna has strong radiation intensity along its end-fire direction.

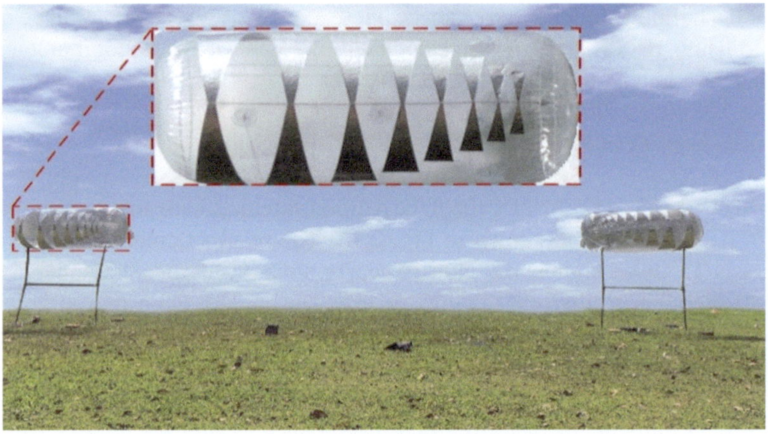

Fig. 2.16 Photograph of the balloon antennas and the experimental setup

Fig. 2.17 Measured and simulated VSWR results of the conformal log-periodic balloon antenna

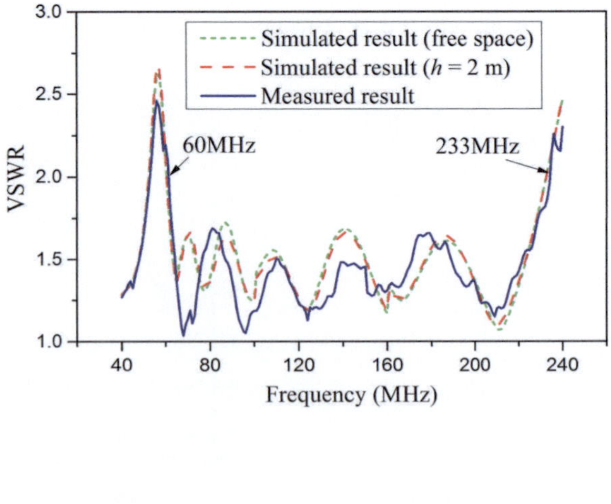

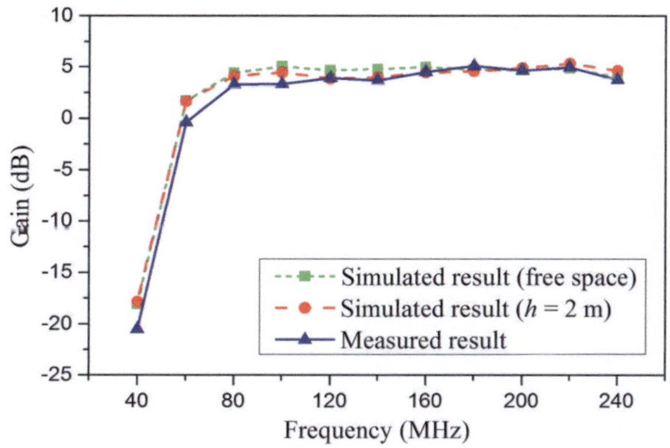

Fig. 2.18 Measured and simulated realized gain results of the conformal log-periodic balloon antenna

2.1.4 Balloon Conformal Array Antenna

Due to their characteristics of low cost, longer hang time and smaller volume after deflation, inflatable balloons are well-suited as aerial base stations for emergency communication applications. Consequently, designing high-performance balloon-conformal array antennas is critical for diverse emergency communication requirements. This chart presents three balloon-conformal array antennas for emergency communication applications, progressively improving beam coverage from directional to half-space beam coverage.

2.2 A Balloon Conformal Array Antenna of Broadened Bandwidth, High Gain and Steerable Directional Beam Fed by a Flexible Phase Shifting Network

This section presents, for the first time, a balloon based conformal array antenna for emergency communication applications in low-range hilly forest scenarios. Fed by a simple, flexible phase shifting network, the antenna features broadened bandwidth, high gain, and steerable directional beam, outperforming other electrically thin conformal array antennas in terms of comprehensive performance. The basic array element is a cylindrically conformal microstrip patch antenna, whose bandwidth is widened by adding parasitic patches loaded by shorting pins for size miniaturization and mutual coupling reduction. Then, a 1×4 conformal subarray with parallel feeding network is designed and optimized for beam sharpening in the elevation plane. To further increase the gain and enable beam steering for full azimuthal coverage, the balloon is surrounded by multiple 1×4 conformal subarrays, and only four of them operate at a time. A flexible phase shifting network is designed for beam steering along the azimuthal direction. To validate the effectiveness of the designs, a prototype was fabricated and measured. Experimental results show the antenna exhibits a gain of around 15 dB and a bandwidth of 74 MHz at the 2.4 GHz band, which has demonstrated the good performance of the balloon conformal array antenna and its huge potential in low-range hilly forest communication applications.

2.2.1 Design of Balloon-Based Conformal Array Antenna

Forest fires, earthquakes, mudslides and other natural disasters have occurred frequently in recent years. Emergency communication plays a vital role in combating these disasters and has garnered significant attention in the past decades. However, many disasters occur in dense forests and hilly regions with adverse channel conditions like severe propagation attenuation and multipath effects, thus posing significant challenges to long-distance emergency communications [3]. The utilization of airborne platforms like unmanned aerial vehicles (UAVs) equipped with high-gain antennas, can be a viable solution [4–6]. However, the UAVs feature relatively high cost and limited endurance restricted by battery capacity [7]. In contrast, inflatable balloons [8–11] have much lower cost, longer hang time and smaller volume after deflation, which is accordingly more suitable for emergency communication applications in the hilly forest environment.

Figure 2.19 shows the emergency communication scenario in the low-range hilly forest environment, where an inflatable balloon serves as the carrier for an aerial base station. During emergency rescue operation, a balloon antenna is utilized to transmit and receive signals to or from the users like rescue personnel. As shown in Fig. 2.19, the elevation angle α is equal to arctan (r/h), where the h and r are the balloon height and the horizontal distance between the user and the balloon, respectively. Considering the required communication distance r (> 500 m) is typically far more than the balloon height h (~ 50 m and can readily be adjustable wherever necessary), the angle α is generally close to 90° for most cases. Accordingly, the maximum beam direction of the presented conformal array antenna (see Fig. 2.20), is designed to lie in the azimuthal plane (i.e., the horizontal direction). Such design can effectively increase the transmission distance as urgently required for low-range hilly forest communication applications, though a blind spot of radius r_1 may exist right below the balloon (see Fig. 2.19). It should be mentioned that for low-range hilly forest regions, the blind spot is very small and can be negligible. For higher mountains, an improved version of the propose antenna (e.g., with a downtilted mainlobe direction) can be designed, and an additional conformal element with wide beam can be located at the balloon bottom to eliminate the blind spot. In addition, a cylindrically conformal assumption is utilized for the central part of the balloon and thus the conformal array antenna, which is reasonable for convenient modeling provided the balloon is sufficiently large.

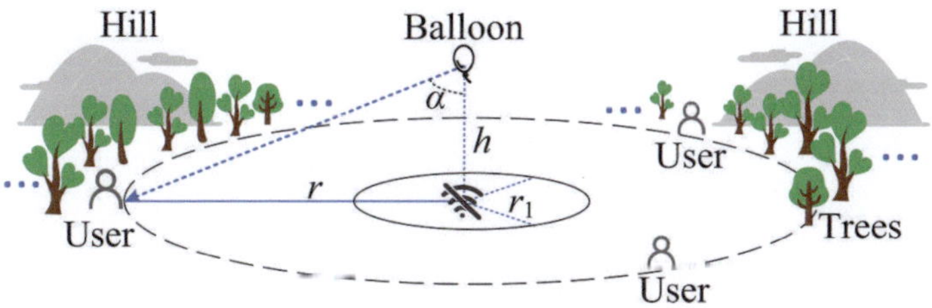

Fig. 2.19 Schematic showing the balloon-based emergency communication application in low-range hilly forest scenarios

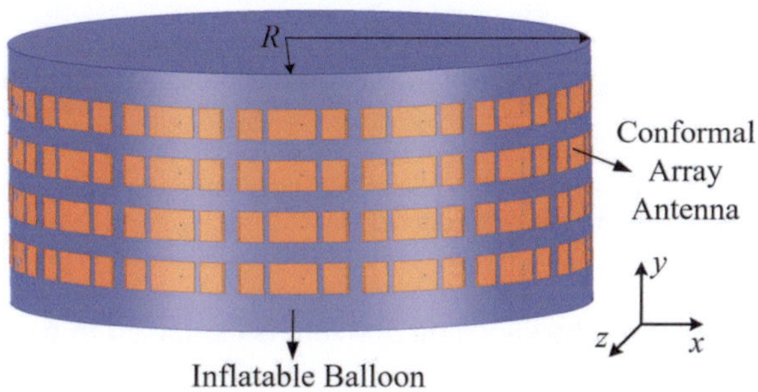

Fig. 2.20 Schematic of the presented balloon-based conformal array antenna

2.2.1.1 Design of Conformal Antenna Element

Benefitting from low profile and easy fabrication, microstrip patch antennas have been widely applied to conformal antenna designs. However, those conformal microstrip antennas typically suffer from extremely narrow bandwidth. Conventional methods like using thick substrate [12] or stacked patches [13, 14] are not feasible for balloon antennas, which would considerably increase bending difficulties due to higher profile and antenna weight thus affecting balloon floatage. Therefore, designing a balloon-based conformal antenna with low profile and wide bandwidth is quite challenging.

Figure 2.21 shows the schematic and planar expanded structure diagram of our designed antenna element. To widen the bandwidth of the conformal antenna with ultra-thin substrate, two parasitic patches are placed nearby both sides of the main patch as depicted in Fig. 2.21a. In this way, the bandwidth of the antenna can be increased by around three times. Unfortunately, the size of the antenna is significantly increased due to the addition of the parasitic patches. To prevent the appearance of grating lobes when forming an array, a series of metalized shorting vias are manufactured on the outer edges of the parasitic patches to reduce their sizes by approximately half. As shown in Fig. 2.21, a piece of 0.5-mm-thick Rogers 5880 material is used as the substrate of the microstrip antenna, which has a relative low dielectric loss of 0.0009 and low dielectric constant of 2.2. Unless otherwise specified, the conformal antenna's shape is a cylinder with radius of 500 mm and the

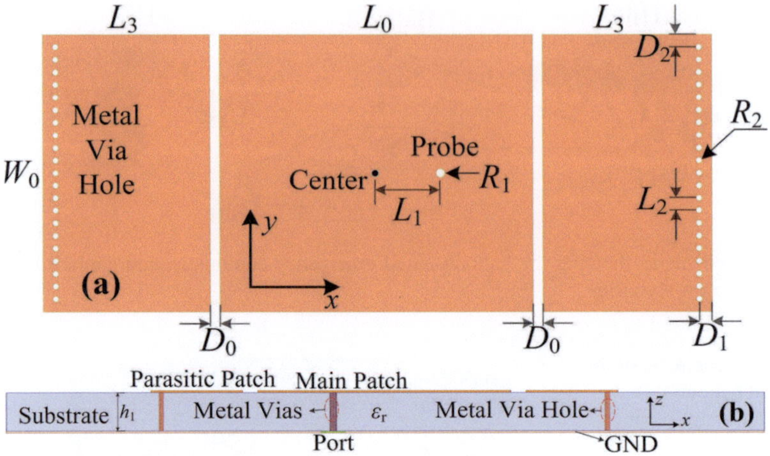

Fig. 2.21 Schematic and planar expanded structure diagram of the conformal antenna element: **a** top-view and **b** cross-sectional view

reason for why choosing 500 mm is explained in part C of Section II. Additionally, all metal structures in the design are copper. Table 2.3 lists the detailed parameters of the designed conformal antenna as shown in Fig. 2.21.

Figure 2.22a shows the simulated reflection coefficients of the presented conformal antenna element, which is compared with that of a planar microstrip patch. Clearly, a new resonant frequency is introduced by the conformal parasitic patches, which has been well adjusted to approach the resonant frequency of the main patch to broaden the bandwidth. It is found that the bandwidth of the conformal element is from 2.38 to 2.43 GHz for $S_{11} < -10$ dB, approximately three times wider than that of the planar patch. Figure 2.22b illustrates the realized gain and radiation efficiency of the conformal element, which are 5.5 dB and 70% at the frequency of 2.4 GHz, respectively. Additionally, the gain changes dramatically at both the high and low frequencies, while the radiation efficiency only drops dramatically near the high frequency. Compared with the planar patch, the conformal element exhibits approximately the same radiation pattern.

Table 2.3 The parameters of the conformal antenna element (unit: mm)

L_0	W_0	L_1	L_2	L_3	h_1
41.5	37.5	8.4	1.7	22.3	0.508

R_1	R_2	D_0	D_1	D_2
0.6	0.4	2.5	1.7	1.35

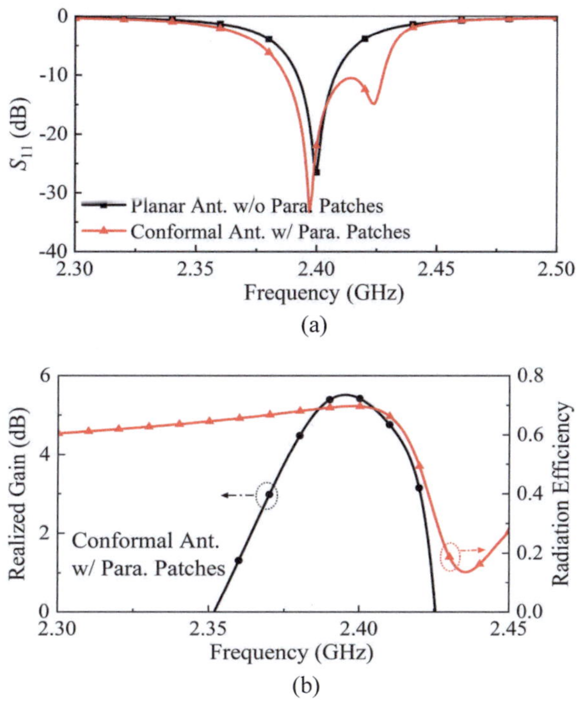

Fig. 2.22 **a** Simulated reflection coefficients of the conformal element and planar patch; **b** simulated realized gain and radiation efficiency of the conformal element

2.2.1.2 Design of Conformal Subarray

For long-range communication in the low-range hilly forest environment, it is necessary to design a conformal array antenna of high gain. Considering that the balloon antenna is preferred to feature a profile as thin as possible, it seems more appropriate to co-planarly design the feeding network and the antenna. However, this not only increases the overall design complexity but also deteriorates the radiation patterns due to the coupling effect between the feeder and the array elements. Accordingly, a double-thin-layer design with shielding ground layer is employed for the radiating elements and feeding network. To avoid radiation performance deterioration with the variation of operating frequency, parallel feeding instead of series feeder is employed for the balloon-based conformal subarray design.

Figure 2.23 shows the planar expanded structure of the conformal subarray. As shown in Fig. 2.23a and b, the subarray consists of four elements aligned along the y-axis with spacing of D_y. Figure 2.23c is a cross-sectional view of the structure, which consists of two thin substrates with a common ground layer in between. The main patch of each subarray element is connected to the microstrip feeding line through a metalized via. As depicted in Fig. 2.23d, the subarray is fed in parallel by a one-to-four T-junction power divider. The symbols Z_{50} and Z_{70} represent the characteristic impedance values of corresponding microstrip lines. Quarter-wavelength transformers are utilized to guarantee good impedance matching. All critical structural parameters are listed in Table 2.4.

Figure 2.24 illustrates the simulated transmission cofficients of the feeding network. To obtain the results, four lumped ports are employed to replace the microstrip patches. It clearly shows that the transmission coefficients are all around − 7 dB over the operating frequency band. Hence, as a quarter power divider with a power distribution ratio of 6 dB, the feeding network has an ohmic loss of around 1 dB. Additionally, the phases of the transmission coefficients are almost identical

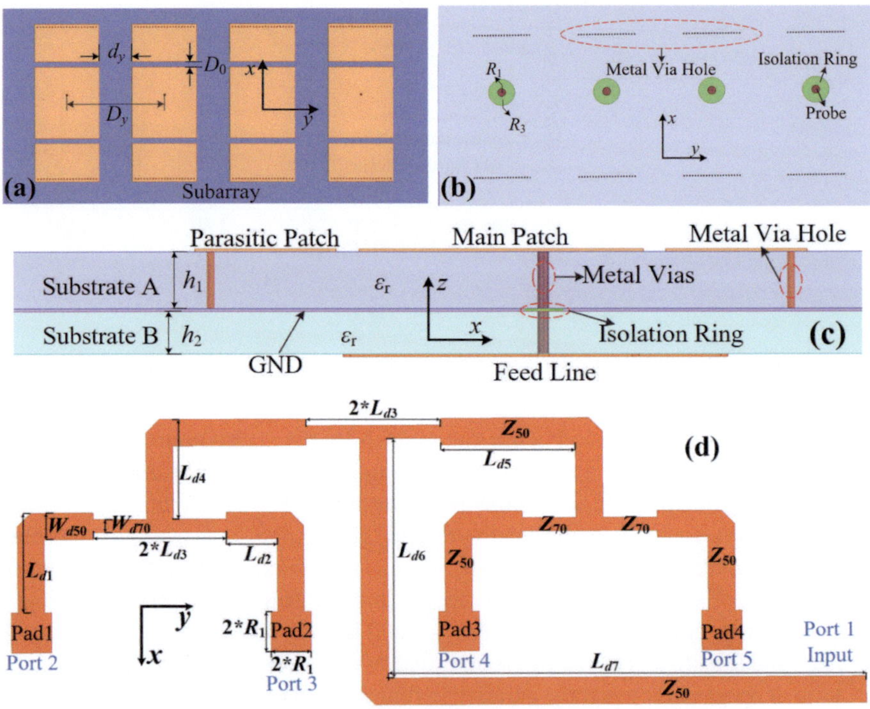

Fig. 2.23 Schematic showing the planar expanded structure of the conformal subarray: **a** top view, **b** ground layer, **c** cross-sectional view and **d** feeding network on the bottom layer

Table 2.4 The parameters of the conformal subarray (units: mm)

d_y	D_y	W_{d50}	W_{d70}	L_{d1}	L_{d2}	L_{d3}
26.25	63.75	0.74	0.42	9.485	8.005	23.5
L_{d4}	L_{d5}	L_{d6}	L_{d7}	R_1	R_3	h_1
9.485	39.88	39.26	133.845	0.6	2.05	0.508

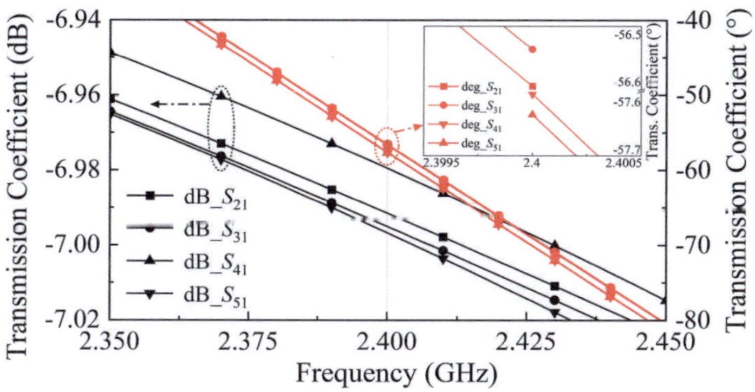

Fig. 2.24 Simulated transmission coefficients of the feeding network

for different output ports, which changes synchronously with the variation of the operating frequency. The low loss and negligible phase difference have shown the good performance of the feeding network.

Figure 2.25a illustrates the simulated reflection coefficients of the conformal element and conformal subarray. It can be found the conformal subarray exhibits a wider operating bandwidth for $S_{11} < -10$ dB of 2.36–2.44 GHz. This mainly results from some extra loss introduced by the feeding network. This is supported by Fig. 2.25b, which indicates that the radiation efficiency of the conformal subarray is 62%, lower than that of the conformal element (~ 70%) at the 2.4-GHz band.

For the conformal subarray, the influence of the distance between the main patch and parasitic patches (i.e. D_0, see Fig. 2.23a) has also been investigated. The primary function of D_0 is to adjust the mutual coupling strength between the conformal main patch and parasitic patches. Figure 2.26 shows the simulated reflection coefficient of the conformal subarray versus the operating frequency for different values of D_0. It is seen that too large value of D_0 (e.g., $D_0 = 2.5$ mm) cannot produce parasitic resonant frequency, while setting D_0 less than 1.0 mm leads to higher return loss in the middle band. Considering error tolerance in practical manufacturing, the value of D_0 is finally chosen to be 1.2 mm.

2.2.1.3 Design of Conformal Subarray

To further enhance the gain and radiation symmetry, four conformal subarrays are combined to form a complete conformal array. The simulation model of the cylindrically conformal array is depicted in Fig. 2.27a, where adjacent subarrays are spaced by an arc length of d_x on a cylinder of radius R. Here, four subarrays are individually fed and beamforming is realized by adjusting their phases. The array element spacing is D_x (approximately $0.74\lambda_0$, where the λ_0 is the wavelength in the free space), which is a bit longer than half wavelength due to the presence of parasitic patches. For convenient description, those conformal subarrays are named C_2, C_1, C_{-1} and C_{-2}

Fig. 2.25 Simulated **a** reflection coefficients and **b** radiation efficiencies of the conformal subarray and conformal element

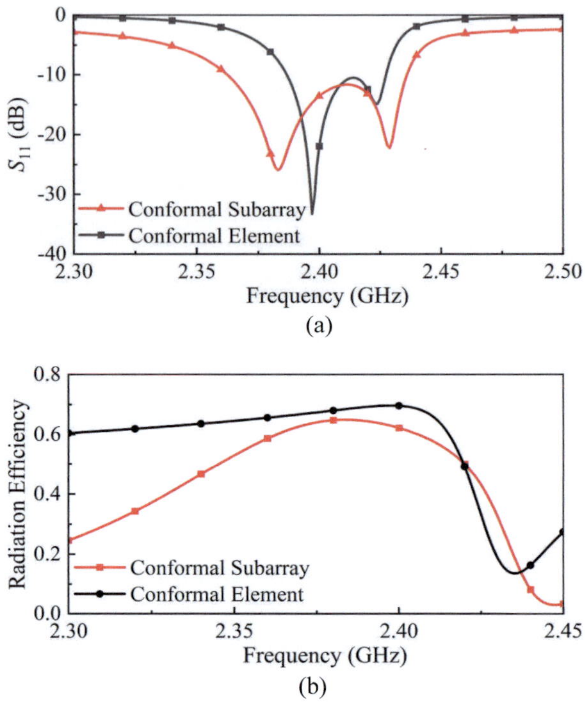

Fig. 2.26 The key parametric study of D_0 for the conformal subarray

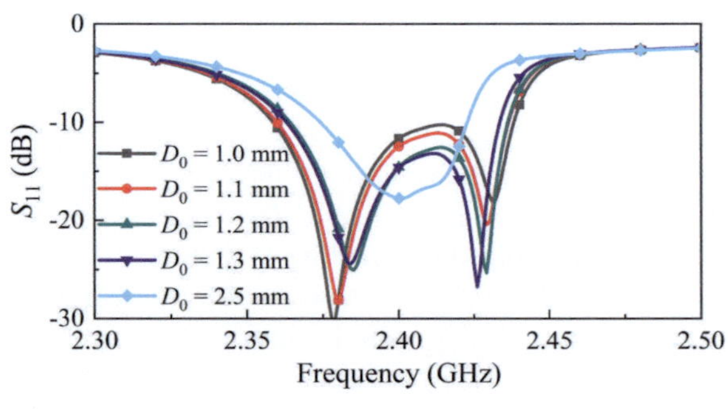

from left to right, respectively. As shown in Fig. 2.27a, the cylindrically conformal array can be considered as a circularly arc array formed by four conformal subarrays.

Figure 2.27b illustrates the beamforming diagram and phase compensation mechanism for the cylindrically conformal array. Assuming the desired maximum radiation direction has a tilted angle of γ with respect to the z-axis, the nearest subarray to the wave front, C_{-1}, serves as the reference. The required phase compensation for the conformal subarrays C_2, C_1, C_{-1} and C_{-2} can be denoted as

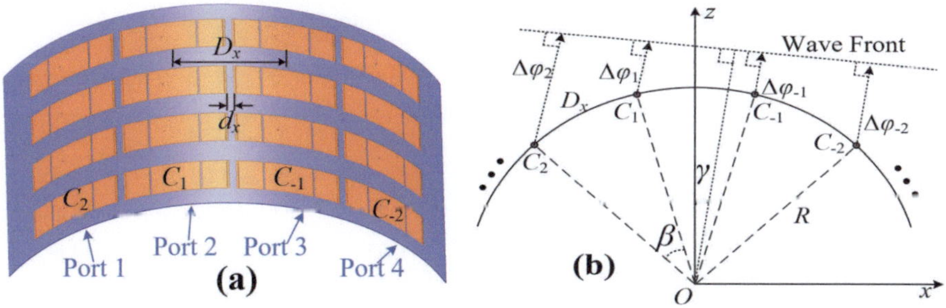

Fig. 2.27 **a** Simulation model and **b** beamforming diagram for the cylindrically conformal array ($D_x = 92.1$ mm and $d_x = 6$ mm)

$$\Delta\varphi_2 = \frac{4\pi}{\lambda_0} R\sin(\beta)\cos\left(\frac{\pi}{2} - \gamma - \frac{\beta}{2}\right) \tag{2.3a}$$

$$\Delta\varphi_1 = \frac{4\pi}{\lambda_0} R\sin\left(\frac{\beta}{2}\right)\sin(\gamma) \tag{2.3b}$$

$$\Delta\varphi_{-1} = 0 \tag{2.3c}$$

$$\Delta\varphi_{-2} = \frac{4\pi}{\lambda_0} R\sin\left(\frac{\beta}{2}\right)\cos\left(\frac{\pi}{2} + \gamma - \beta\right) \tag{2.3d}$$

where β is the angle corresponding to the arc spacing D_x, i.e.,

$$\beta = \frac{D_x}{R} = \frac{L_0 + d_x + 2 \times (L_3 + D_0)}{R} \tag{2.4}$$

Particularly, when the desired wave front is perpendicular to the z-axis (i.e., $\gamma = 0$), the calculated phase compensation is $\Delta\varphi_1 = \Delta\varphi_{-1} = 0$ and $\Delta\varphi_2 = \Delta\varphi_{-2}$.

Table 2.5 compares the radiation performance of the conformal array with different cylindrical radii, including the realized gain, HPBW and maximum side-lobe level (MSLL) in both *xz* and *yz* planes. The results indicate that all performance specs are improved with increasing the radius. For emergency communications in the low-range hilly forest environment, a MSLL of lower than − 12 dB is required for reducing multipath effects and interferences. Larger cylindrical radius can bring lower MSLL, but also lead to more subarrays and thus higher cost of system. Therefore, we choose the cylindrical radius to be 500 mm, which makes a good compromise between performance and cost.

Figure 2.28 depicts the simulated current distribution on the cylindrically conformal array, which clearly indicates the desired amplitude and phase distribution. The four elements on every conformal subarray have equal-amplitude and in-phase surface currents. Due to the structural symmetry, the current distribution is also symmetrical for the conformal array, which leads to a maximum radiation toward

Table 2.5 Simulated performance of the conformal array with different cylindrical radii

Radius (mm)	Realized gain (dB)	HPBW (°)		MSLL (dB)	
		xz plane	*yz* plane	*xz* plane	*yz* plane
300	14.5	19.6	24.8	– 9.57	– 14.3
400	14.7	19.54	24.98	– 11	– 14.45
500	15	19.3	25.15	– 12.2	– 14.5
600	15.1	19.16	25.45	– 12.05	– 14.6
Planar (∞)	16	16.42	25.14	– 12.62	– 15

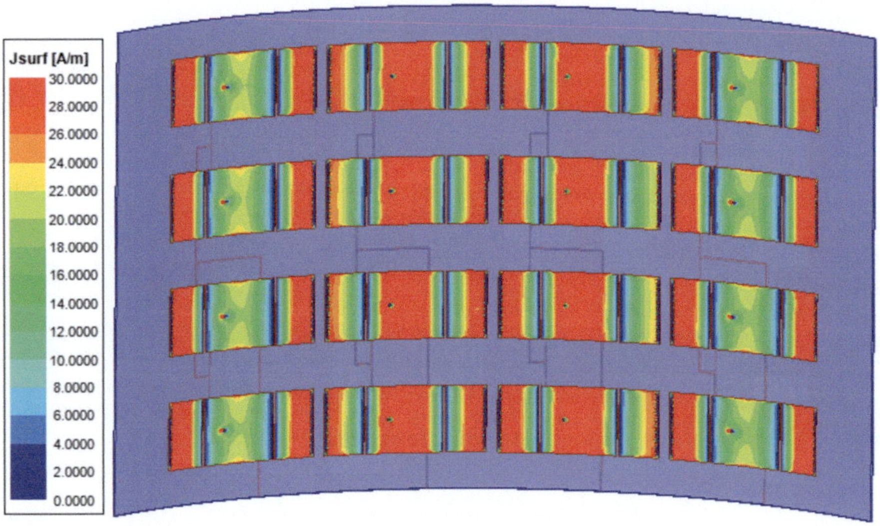

Fig. 2.28 Simulated current distribution on the cylindrically conformal array

the z axis. Furthermore, the current phases of the middle two subarrays lag behind as compared with those on the outer ones. Figure 2.29 illustrates the mutual coupling between the conformal subarrays. The maximum mutual coupling is weaker than – 20 dB, which indicates good isolation characteristics.

2.2.2 Design of Flexible Phase Shifting Network

For emergency communication applications, phase shifting network is necessary for dynamic beam steering and full azimuthal signal coverage. Due to the high cost of digital phase shifters and the complexity of feeding networks [15], currently reported phase shifting network designs can hardly be converted to conformal circuits. To tackle the challenge, we propose a flexible phase shifting network that is conformal

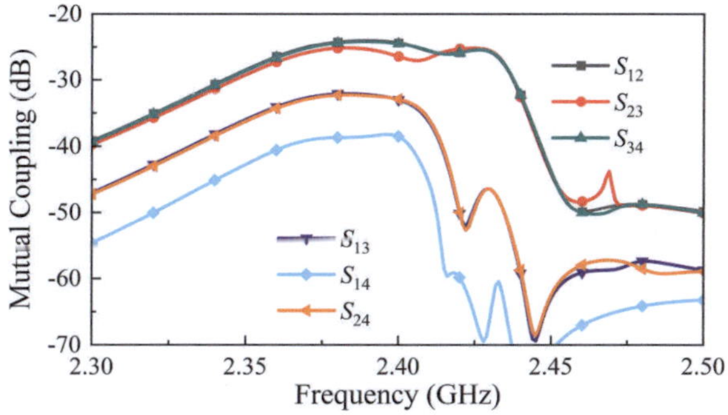

Fig. 2.29 Simulated mutual coupling between the conformal subarrays

with the balloon, which combines the techniques of analog phase shifting and sequential beam switching for significantly facilitating conformal implementation.

Figure 2.30 illustrates the planar expanded structures of the flexible phase shifting network, where each branch is to be connected to a conformal subarray. As depicted in Fig. 2.30a, 50-Ω microstrip assembly line is utilized to feed all branches. For beam forming and steering, only four branches are switched on at a time while remaining ones turned off. As shown in Fig. 2.27b and Table 2.5, the minimum beam steering angle β is around 11° and thus less than the HPBW (19.3°) of the conformal array with radius of 500 mm, which can guarantee a full azimuthal coverage while maintaining a high gain. As shown in Fig. 2.30a, the branch spacing L_{50} is equal to one guided wavelength. Considering the required input impedance is 200 Ω at each branch connection, a 100-Ω quarter-wavelength impedance transformer is employed for impedance matching. Additionally, metalized shorting via holes are added to the end of the assembly line, which are located at a quarter of guided wavelength from the last branch. This ensures that the assembly line is equivalent to a ground for direct current (DC) biasing.

To reduce the fabrication cost, a simplified phase shifting network with six branches is designed. As shown in Fig. 2.30a and b, each branch is equipped with four PIN diodes, controlled by two DC biasing lines to achieve on/off switching and desired phase compensation. Specifically, PIN1 is responsible for connecting or disconnecting the branch to the assembly line, while PIN2, PIN3 and PIN4 allow analog phase shifting and switching between two alternative statuses for the ahead phase. Additionally, each branch integrates two inductors and one capacitor to realize radio-frequency (RF) and DC isolation. The structural parameters shown in Fig. 2.30 are listed in Table 2.6.

To facilitate the analysis, three operation cases are considered, named Case-1 (the branches 1, 2, 3 and 4 active), Case-2 (the branches 2, 3, 4 and 5 active) and Case-3 (the branches 3, 4, 5 and 6 active), respectively. Figure 2.31 illustrates the simulated results of S_{11} for those three cases, which indicates the operating bandwidth well covers the operating frequency range of the conformal array.

Figure 2.32 shows the simulated amplitudes of the transmission coefficients for different branches for three operating cases. Taking Fig. 2.32a as an example, the

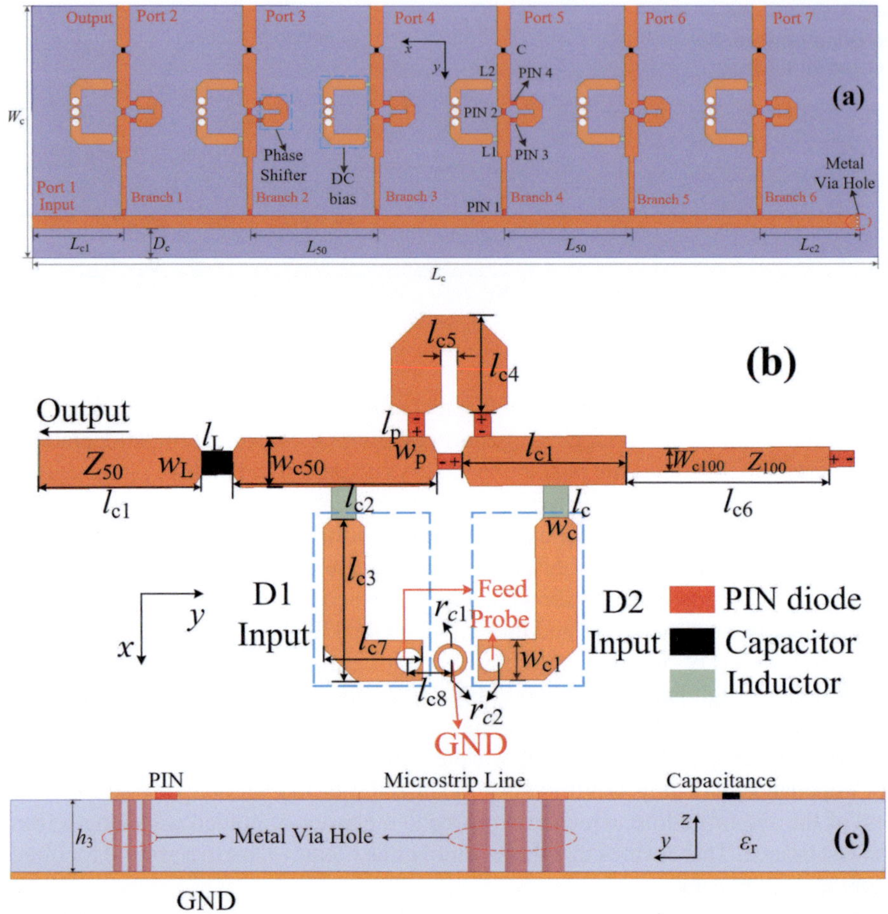

Fig. 2.30 Planar expanded structures of the flexible phase shifting network: **a** overall structure, **b** individual branch details and **c** cross-sectional view

Table 2.6 The parameters of the flexible phase shifting network (unit: mm)

L_c	W_c	L_{c1}	L_{c2}	L_{50}	D_c	l_{c1}	l_{c2}	l_{c3}
500	70	20.725	23.875	90.4	10	10	13.7125	10

l_{c4}	l_{c5}	l_{c6}	l_{c7}	l_{c8}	w_{c50}	w_{c100}	l_L	w_L
4.6	1	23.5875	6.12172	2.54	3.05	0.85	0.3	0.5

l_p	w_p	l_c	w_c	w_{c1}	r_{c1}	r_{c2}	h_3
1.2	0.8	0.3	0.5	2.5	1.27	0.9652	1

Fig. 2.31 Simulated reflection coefficients of the flexible phase shifting network for three cases

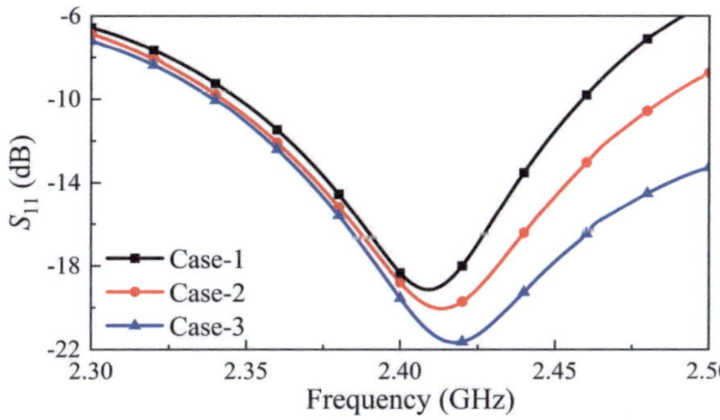

left four branches are on and the other two branches are off. It is clearly seen that the transmission coefficients of the phase shifting network are around 7.2 dB for all active branches, a sum of the one-to-four power distribution loss (6 dB) and the ohmic loss (~ 1.2 dB) due to the finite conductivity of metalized layers and the nonideal characteristics of the PIN diodes used. Furthermore, more PIN diodes used in the phase shifter would bring higher loss, resulting in minor amplitude differences between different branches. Nevertheless, those differences are negligible, which have little influences on the radiation pattern and realized gain of the conformal array. Using better PIN diodes can lead to higher performance. Additionally, the isolation between the Port 1 and the Port 6 or 7 is approximately 14 dB at the 2.4-GHz band.

Figure 2.33 shows the simulated phases of the transmission coefficients for different branches. Considering the maximum radiation direction is designed toward the z axis, the required phase compensation can be calculated as $\Delta\varphi_1 = \Delta\varphi_{-1} = 0$ and $\Delta\varphi_2 = \Delta\varphi_{-2} = 51.1°$ for a cylindrical radius $R = 500$ mm. This means that the current phases on outer two subarrays should be 51.1° ahead of those on the middle two subarrays. Taking Fig. 2.33a as an example (i.e., Case-1), the current phase on Branch 1 (or 2) are approximately equal to that on Branch 4 (or 3) at the central frequency of 2.4 GHz. Moreover, the phase difference is around 51.3°, which is very close to the value as calculated using Eqs. (3.1) and (3.2). Figure 2.33b and c also show the similar results for the other cases.

Figure 2.34 shows the simulated current diagram on the phase shifting network for three cases. Taking Case-1 as an example, as shown in Fig. 2.34a, the electric

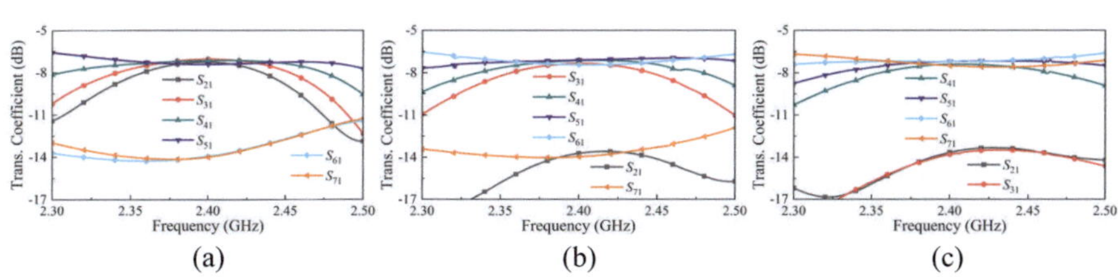

Fig. 2.32 Simulated amplitudes of the transmission coefficients for different branches for **a** Case-1, **b** Case-2 and **c** Case-3

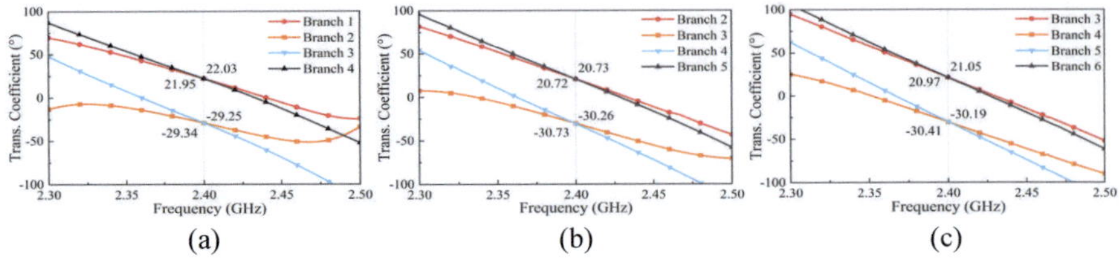

Fig. 2.33 Simulated phases of the transmission coefficients for different branches for **a** Case-1, **b** Case-2 and **c** Case-3

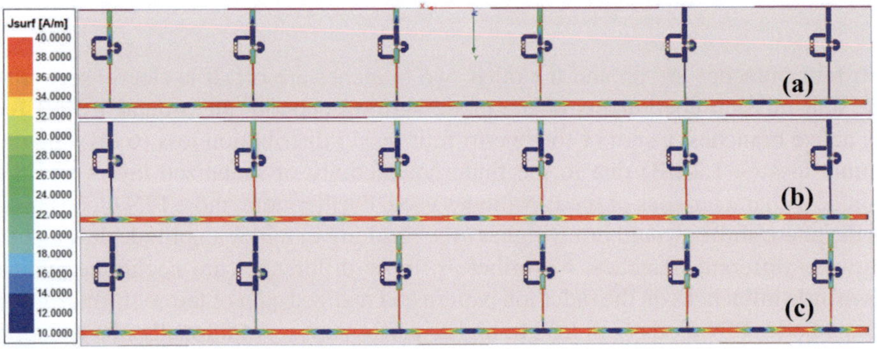

Fig. 2.34 Simulated current diagram on the flexible phase shifting network for **a** Case-1, **b** Case-2 and **c** Case-3

currents on the left four branches are strong and uniform while nearly disappearing on the right two branches. Furthermore, the current phases on the branches 1 and 4 are significantly ahead of the branches 2 and 3. The other two cases are similar, which fulfill the design objectives of phase shifting and beam steering.

2.2.3 *Experimental Measurements and Analysis*

2.2.3.1 Experimental Details and Measurements of the Conformal Subarray

Figure 2.35 shows the photographs of the fabricated conformal array and phase shifting network as well as the experimental setup in the anechoic chamber. To ensure good conformal characteristics with the surface of the balloon, the F_4BM220 dielectric substrate was chosen due to its relatively high flexibility, low relative permittivity and low dielectric loss. During prototype manufacturing, all metallized parts are copper with a thickness of 17 μm. The fabricated conformal array and phase shifting network have expanded planar dimensions of $4.96\lambda_0 \times 2.144\lambda_0 \times 0.0072\lambda_0$ and $4\lambda_0$

Fig. 2.35 Photographs of the **a** conformal array, **b** phase shifting network and **c** measurement setup in the anechoic chamber

$\times$ $0.56\lambda_0$ $\times$ $0.008\lambda_0$, respectively. Also, a cylindrical balloon made of polyvinyl chloride with a radius of 500 mm was manufactured. As depicted in Fig. 2.35c, benefitting from their ultrathin profiles and excellent flexibility, the conformal array and phase shifting network are easily attached with the inflatable balloon.

As for the phase shifting network, choosing proper electronic components is critical before experiments. The circuit employs chip capacitors and inductors in 0402 packages due to their low cost and excellent performance, with capacitance and inductance of 100 nF and 56 nH, respectively. For the selection of PIN diodes, the forward resistance and open-circuit capacitance can significantly impact circuit performance. The BAR64-02 V silicon PIN diodes are chosen due to their relatively low capacitance and forward resistance, which are 0.17 pF and 1 Ω at 100-mA forward current, respectively.

Figure 2.36 shows the simulated and measured scattering parameters of the conformal subarrays. As depicted in Fig. 2.36a, the measured results indicate that the operating bandwidth for $S_{11} < -10$ dB is from 2.356 to 2.43 GHz, which agree well with the simulated bandwidth of 2.36–2.44 GHz. This minor difference may be caused by the dielectric constant error of the substrate or manufacturing tolerance. Figure 2.36b depicts the mutual coupling between the conformal subarrays,

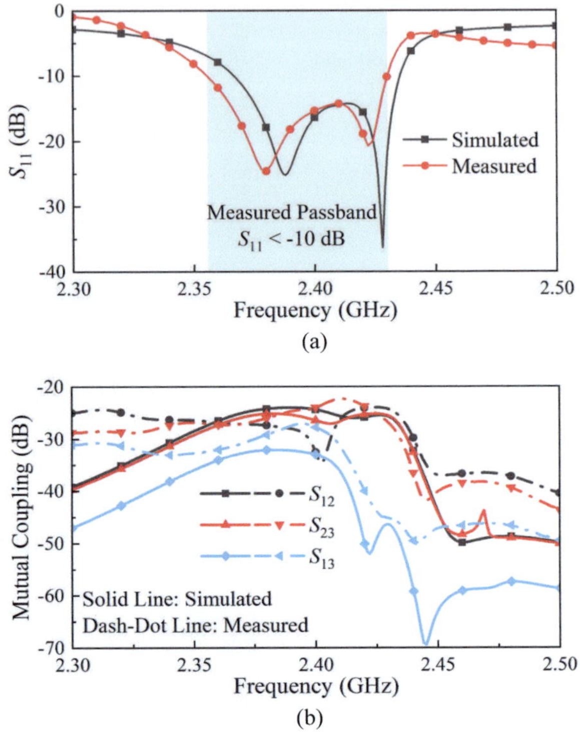

Fig. 2.36 Measured and simulated scattering parameters of the conformal subarrays: **a** S_{11} and **b** mutual coupling

where both simulation and experimental results consistently indicate that the mutual coupling is weaker than − 20 dB.

Figure 2.37 shows the simulated and measured radiation patterns of the conformal subarray. The measured and simulated results are in well agreement in both *xz* and *yz* planes with a peak gain of 9.6 dB. For an isolated subarray, the radiation pattern in the *xz* plane is symmetrical. However, spatial coupling slightly distorts the radiation pattern when the subarray is located within an array. In the *yz* plane, the radiation patterns of the conformal subarray exhibit excellent symmetry, which is exactly as same as that of the isolated subarray.

2.2.3.2 Measurements of the Flexible Phase Shifting Network

For characterizing the performance of the flexible phase shifting network, the DC biasing voltages should be reasonably configured for controlling the on/off states of the branches and achieving required phase shifting compensation. For convenience, the bias voltage for each branch is named V_{ab}, where a and b are the branch number and DC input number (i.e., $a = 1, 2, \ldots, 6$; $b = 1, 2$), respectively. For example, V_{32} denotes the voltage at the lower DC input pad of the third branch. Detailed

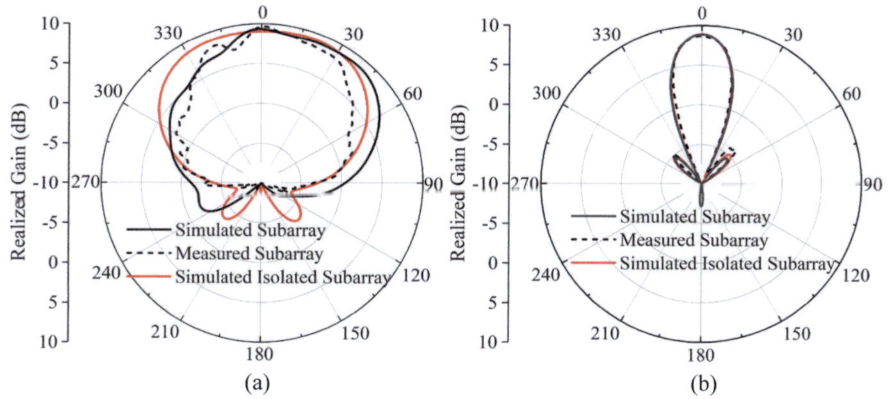

Fig. 2.37 Measured and simulated radiation patterns of the conformal subarray. **a** *xz*-plane. **b** *yz*-plane

configuration of V_{ab} for three cases are listed in Table 2.7, where V_F is the forward voltage of the PIN diode for normal operation. Additionally, the forward current I_F was set to 100 mA for reducing the ohmic loss caused by the PIN diode. During measurement, a RIGOL DP832 DC source was utilized to supply DC biasing for the network.

Figure 2.38 shows the measured and simulated scattering parameters of the flexible phase shifting network. As depicted in Fig. 2.38a, although there is slight deformation in S_{11} curves for all three cases, a wide passband is guaranteed. Taking Case-1 as an example, the transmission coefficients are illustrated in Fig. 2.38b and c. The experimental results reveal a minor degradation of the amplitude, which could be attributed to various factors such as soldering losses and fabrication errors. Nevertheless, such minor degradation would not considerably impact the antenna's performance. Additionally, Fig. 2.38d depicts the phases of transmission coefficients for Case-1. On the whole, all the measured results are in reasonably agreement with the simulated results.

Table 2.7 Configuration of biasing voltages for the phase shifting network ($V_F = 1$ V)

	V_{11}	V_{21}	V_{31}	V_{41}	V_{51}	V_{61}	V_{12}	V_{22}	V_{32}	V_{42}	V_{52}	V_{62}
Case-1	GND	$3V_F$	$3V_F$	GND	–	–	V_F	V_F	V_F	V_F	GND	GND
Case-2	–	GND	$3V_F$	$3V_F$	GND	–	GND	V_F	V_F	V_F	V_F	GND
Case-3	–	–	GND	$3V_F$	$3V_F$	GND	GND	GND	V_F	V_F	V_F	V_F

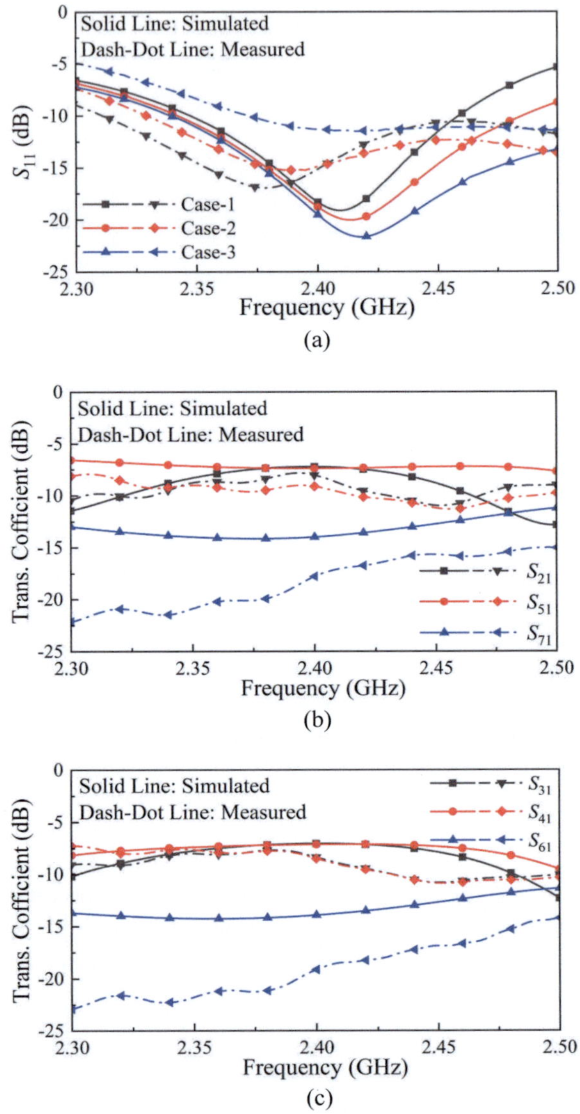

Fig. 2.38 Measured and simulated scattering parameters of the flexible phase shifting network: **a** S_{11} for different cases; **b** and **c** the amplitudes of transmission coefficients for Case-1; **d** the phases of transmission coefficients for Case-1

2.2.3.3 Combination of the Conformal Array and the Phase Shifting Network

Figure 2.39a shows the measured and simulated reflection coefficients for the antenna system combining the conformal array and the phase shifting network. Full-wave analysis of the combined antenna system is very time-consuming and computationally effortful. Hence, the simulated reflection coefficient as shown in Fig. 2.39a are obtained using the software high frequency structure simulator (HFSS) and advanced

Fig. 2.38 (continued)

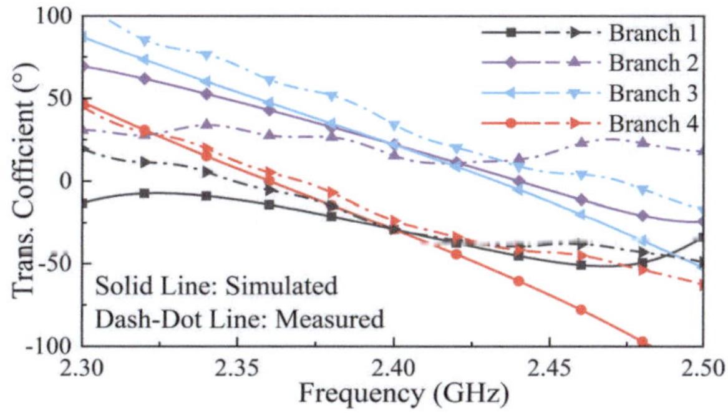

(d)

design system (ADS). Specifically, all scattering parameters of the conformal array and the phase shifting network are obtained separately in the HFSS, and then a combined network is modelled in ADS and simulated. As depicted in Fig. 2.39a, the measured and simulated S_{11} curves have the similar shape, though there is a bit frequency offset caused by dielectric constant or manufacturing errors. In addition, a wider measured operating bandwidth is observed as compared with simulation, which mainly results from various losses in experiments. Figure 2.39b depicts the measured and simulated gain versus the operating frequency. The simulated gain curve refers to the conformal array only, so its values are slightly higher than the measured gain of the combined antenna system. Good consistence between the measured and simulated results have confirmed the effectiveness of our design.

Figure 2.41 shows the measured and simulated radiation patterns of the combined antenna system for the three cases at 2.4 GHz. To visualize the beam steering performance with corresponding scanning losses, a Cartesian format is employed. The proposed antenna system employs a phase shifting network for beam steering along the azimuthal direction, while the radiation patterns in the yz plane remain unchanged. Thus, it suffices to show the radiation pattern of the yz plane for one case. Figure 2.40 clearly depicts the steerable beams under the control of the phase shifting network, which results in a beam direction rotation by β (about 11°) for adjacent cases. Additionally, since the minimum beam steering angle β is less than the HPBW (19.3°) of the conformal array, which ensures a full azimuthal coverage while maintaining a high gain. The measured results indicate that the gain loss associated with beam steering is around 0.2 dB. The measured HPBW of the antenna system in the *xz* and *yz* planes are 17° and 23°, respectively, which are slightly narrower than the simulated results (19.3° and 25.15°). This minor discrepancy can be attributed to manufacturing and measurement errors.

Figure 2.41a–d shows the radiation patterns in the *xz* and *yz* planes at 2.42 GHz and 2.36 GHz, respectively. It can be found the simulated and measured results are in good agreement at the lower band, while the measured gain at the high frequency is slightly lower than the simulated one. This is mainly due to the frequency offset

Fig. 2.39 Measured and simulated results for the combination of the conformal array and the phase shifting network: **a** S_{11} and **b** realized gain

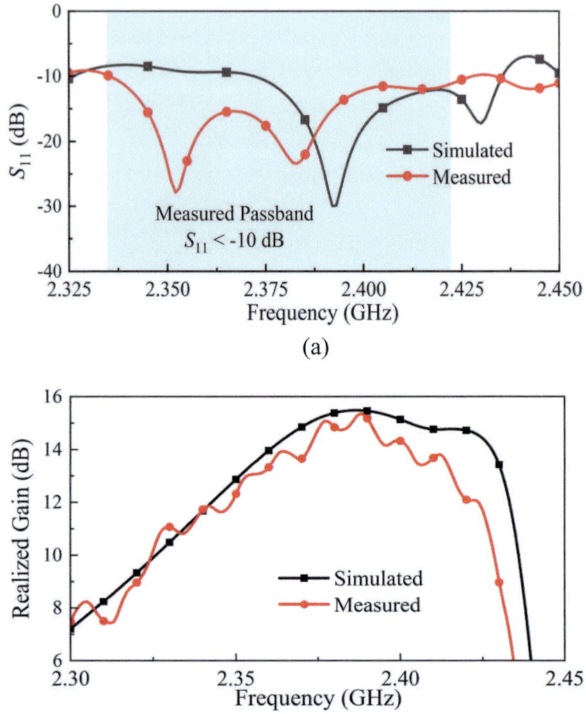

Fig. 2.40 The measured and simulated radiation patterns of the combined antenna system for the three cases

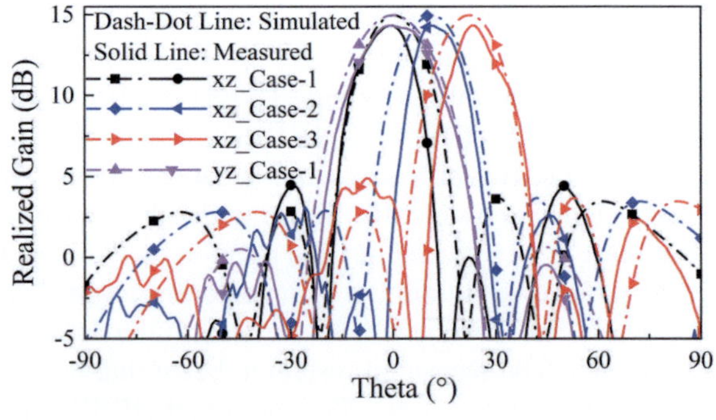

toward lower frequencies for the manufactured prototype, as depicted in Fig. 2.39a. Additionally, as shown in Fig. 2.39b, the gain near the high frequency drops quicker than that at the lower band, resulting a slightly lower measured gain than simulation.

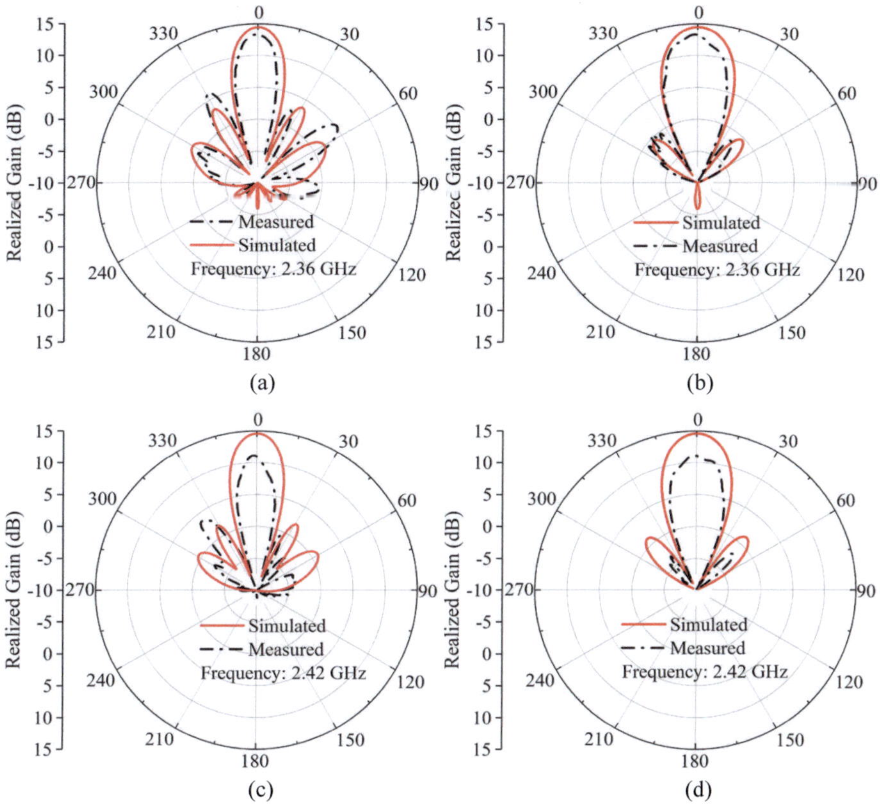

Fig. 2.41 Measured and simulated radiation patterns of the combined antenna system: **a** *xz* and **b** *yz* planes for Case-1 and 2.36 GHz; **c** *xz* and **d** *yz* planes for Case-1 at 2.42 GHz

2.3 A Balloon Conformal Heterogeneous Array Antenna with Ultrawide Bandwidth and Half-Space Beam Coverage

This section presents a balloon-based conformal heterogeneous array antenna with ultrawide bandwidth and half space beam coverage for emergency communication applications (as depicted in Fig. 2.40). The antenna consists of two types of cylindrically conformal arrays, each of which contains nine broadside or endfire sub arrays. By virtue of 1-D full-azimuthal beam switching and combining the radiation patterns of those two arrays, one yields an entire half-space beam coverage. To increase the bandwidth and gain while maintaining low profiles, the broadside subarray utilizes 1 × 4 dumbbell-shaped full-wavelength dipoles cross-fed in series and backed by a miniaturized artificial magnetic conductor (AMC); the endfire subarray employs 2 × 3 size tapered half-wavelength dipoles with optimized spacing to achieve the desired radiation characteristics. To validate the designs, the prototypes of two conformal subarrays with one shared aperture were fabricated and measured. Experimental results have indicated the proposed conformal heterogeneous array exhibits an overlapped bandwidth of 720 MHz and 360° × 108° beam coverage at the 2.4-GHz band, which shows its superior performance and its huge potential in the mountainous and dense-forest communication applications.

2.3.1 Design and Analysis of the Balloon-Based Conformal Heterogeneous Array Antenna

2.3.1.1 Design of Balloon-Based Conformal Heterogeneous Array Arrangement

Figure 2.42 shows the structural geometry of the proposed balloon-based conformal heterogeneous array antenna. It is worth mentioning that a cylindrical conformal assumption is utilized for the localized zone of the balloon and thus the array antenna, which is reasonable for convenient modeling as long as the balloon is sufficiently large. As illustrated in the Fig. 2.42, two types of conformal arrays are closely spaced and uniformly arranged on the surface of a cylindrical balloon. For the conformal arrays, the numbers and spacing of their subarrays depend on the beamwidth of the antenna and the radius of the balloon. The effects of the balloon radius on antenna performance is discussed in Section II-C and II-D, and the fabricated balloon prototype has a radius of 500 mm (or $4\lambda_0$, where λ_0 is the wavelength in the free space at 2.4 GHz). It is clearly seen from Fig. 2.42 that each conformal array contains nine broadside subarrays (i.e., the Subarray A) or nine endfire subarrays (i.e., the Subarray B), all of which are cylindrically conformal radiators of ultrawide operating bandwidth. Featuring a high gain and a highly directional radiating beam, the Subarray A is mainly utilized to communicate with remote users or ones at similar altitudes (e.g., air to air communications between the balloon-based aerial station and other airborne platforms, see Fig. 2.42). In comparison, the Subarray B with moderate gain features an exactly downward fan-shaped endfire beam, which covers the most part of lower half space and thus is utilized for relatively short-range air-to-ground communications for users such as the rescue personnel and vehicles (Fig. 2.43).

Figure 2.44a and b show the switchable radiation patterns in the azimuthal plane for the conformal Subarrays A and B, respectively. It should be emphasized that for achieving structural compactness, convenient feeding and easy assembly, the Subarrays A and B are designed to have a shared aperture. Considering the 3-dB azimuthal beamwidth of Subarray A is design for achieving high gain, only nine subarrays spaced by 40° are necessary to cover all horizontal directions. On the other hand, the Subarray B features a fan-shaped endfire radiation pattern with quasi-bi-directional rather than unidirectional beam projection in the azimuthal plane. Taking into consideration the difference in beam projection between the Subarrays A and B,

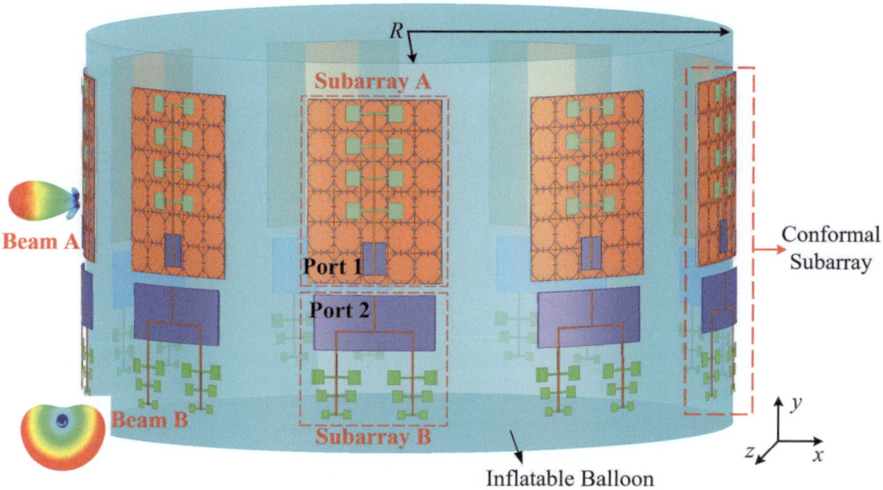

Fig. 2.42 Structural geometry of the balloon-based conformal heterogeneous array antenna

Fig. 2.43 Schematic showing the balloon-based emergency communication scenario in a mountainous and dense-forest environment

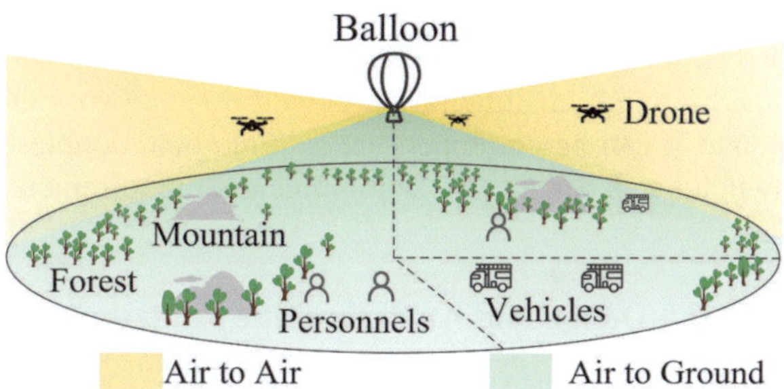

as shown in Fig. 2.44b, nine subarrays spaced by 20° are designed for the conformal endfire array to achieve full azimuthal beam coverage. Figure 2.44c shows the beam combination in the elevational plane for the conformal subarrays A and B. Obviously, the Subarray B has a much wider (more than 170°) beam in the elevational plane as compared with that in the azimuthal plane, which can be well jointed with the radiation beam of the Subarray A for practical wireless applications. As shown in Fig. 2.44c, for short-range air-to-ground communications, the Subarray B is utilized to achieve a wide beam coverage when the users are located within the angular range from 113° to 247°. Otherwise, the Subarray A is used to achieve long-range air-to-ground or air-to-air communications. Therefore, a full beam coverage can be achieved for the lower half space based on such array arrangement.

It should be mentioned that although wide beam coverage can also be achieved using an electrically small antenna (i.e., the dipole) with omni-directional pattern, the maximum gain of around 2.1 dB is too low to compensate considerable link propagation loss in harsh mountainous and dense-forest environment. The proposed

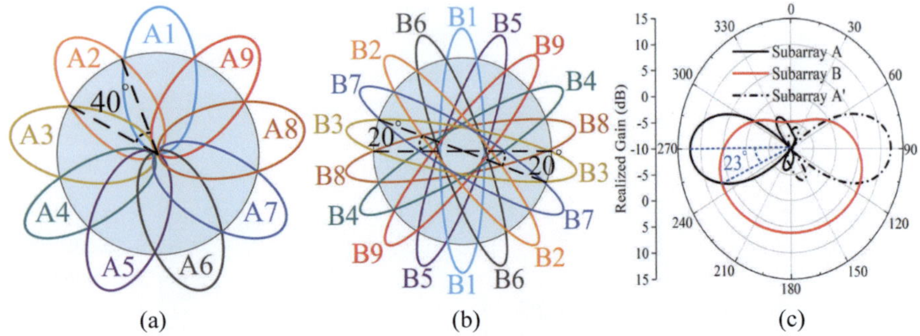

Fig. 2.44 Switchable radiation patterns in the azimuthal plane for the conformal **a** Subarray A and **b** Subarray B, respectively; **c** beam combination in the elevational plane for those two conformal subarrays

balloon conformal heterogeneous array antenna features half-space beam coverage, ultrawide bandwidth and moderate gain (of around 12 dB in maximum radiation direction), thus very suitable for large-capacity wideband air-to-air and air-to-ground emergency communications in the mountainous and dense-forest environment. In practical applications, electrically controllable beam switching enables rapid beam steering while maintaining relatively low network complicity. Also, the subarrays A and B can be combined via a single-pole double-throw switch to achieve beam switching as per specific communication requirement.

2.3.1.2 Design and Analysis of the Miniaturized AMC

To start with the design of the conformal Subarray A for balloon-based communication applications, a low-profile, lightweight and wideband AMC is designed first. AMCs are widely used in antenna designs due to its in-phase reflection characteristics, which can effectively reduce the backlobe and thus mutual interferences between array elements [16, 17]. For balloon antenna designs, currently reported patch-type AMCs are more preferred due to its relatively simple single-layer structure. However, the large dimensions of the unit cells result in that the AMCs are sensitive to the variation of the incident angle, thus bringing challenges to conformal antenna designs. In addition, a relatively thick dielectric substrate is often required for the AMC to increase the operating bandwidth. However, this is unfavorable for balloon antenna designs. Especially when designing a conformal array antenna, an AMC with large size needs to be utilized as the reflecting plane, whose weight may affect the floating capability of the inflatable balloon. Therefore, the design of a lightweight wideband AMC with miniaturized unit size is of significant importance for designing balloon-based conformal array antennas.

Figure 2.45 shows the evolutionary design process of the proposed AMC unit cell. As depicted in Fig. 2.45a–c, G I is an initial structure with a circle patch and ground on the top and bottom layers of the substrate, respectively. G II is an improved version of G I, where four slots are cut symmetrically to prolong the electric current path for size reduction. To further reduce the size of the unit cell, four arrow-shaped arms are added symmetrically to four corners of G II, which creates G III. In addition, the cross-sectional view of the proposed AMC structure is depicted in Fig. 2.45d. To reduce the weight and improve the bandwidth of the AMC, a piece of foam is inserted between the top and bottom substrates, which are fabricated using two pieces of 0.25-mm-thick Rogers 5880 (relative permittivity: $\varepsilon_{r2} = 2.2$) materials. The foam is very bendable and much lighter than the substrates, thus beneficial for implementing lightweight and flexible AMC. Since the dielectric constant of the foam is very low, approximately as same as that of the air, utilizing the foam brings lower equivalent dielectric constant as compared with that of the pure substrate, thus helpful for widening the bandwidth. Although the effects of the foam and air gap are basically the same on electrical performance, the former can provide much better mechanical support for the overall conformal structure with stable and accurate dimensions against bending.

The simulated reflection phases and current distribution of those three AMC structures are illustrated in Fig. 2.46. It can be clearly seen that the zero-degree reflection phase moves to lower frequency during evolutionary design process of the proposed AMC unit cell. This is because the electrical current path is prolonged by adding additional slots and arms. In particular, the four arrow-shaped arms have more pronounced effects on lengthening the current path than those four slots. This also explains why the zero-degree phase reflection frequency for Generation III is much lower than that for Generation II.

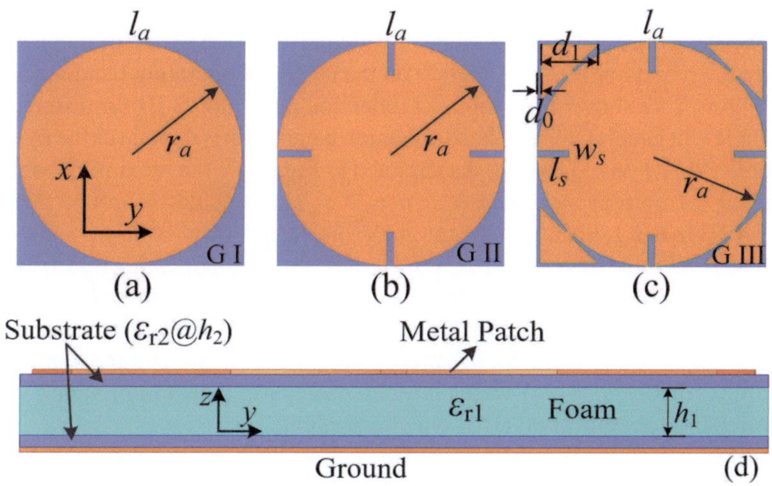

Fig. 2.45 Evolution of the AMC unit cell: **a** traditional circular patch (Generation I); **b** slot loaded circular patch (Generation II); **c** top and **d** cross-sectional views of the proposed AMC unit cell (Generation III). Size parameters: $r_a = 16.6$, $l_a = 33.4$, $d_0 = 0.1$, $d_1 = 9.1$, $l_s = 4$, $w_s = 1$, $h_1 = 2.5$, $h_2 = 0.25$, $\varepsilon_{r1} = 1$, $\varepsilon_{r2} = 2.2$ (the units are mm except for ε_{r1} and ε_{r2})

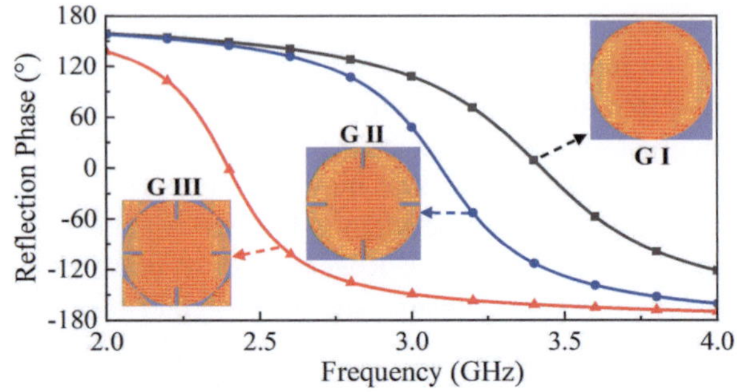

Fig. 2.46 Simulated reflection phases and current distribution of three AMC structures as shown in Fig. 2.45. All AMC structures have the same cross-sectional view as shown in Fig. 2.45d

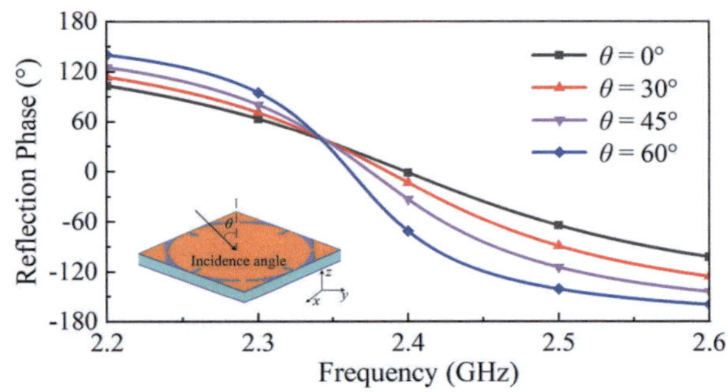

Fig. 2.47 Simulated reflection phases of the proposed AMC structure for different incident angles

To better demonstrate the superiority of our proposed miniaturized AMC, its advantages for designing the conformal antenna is also investigated. Since the Floquet port in the electromagnetic simulation software only support planar periodic models, we study the effects of incident angle θ on the available operating bandwidth of the AMC. Figure 2.47 shows the simulated reflection phases of G III for different incidence angles. In general, the in-phase reflection bandwidth is defined as the frequency range of [$-90°$, $+90°$] [18, 19]. As depicted in Fig. 2.47, the maximum bandwidth is obtained for normal incidence (i.e., $\theta = 0°$), which achieves around 330 MHz at the 2.4 GHz band. For the incident angle increasing from 0° to 60°, the central frequency is slightly lowered to 2.36 GHz and the in-phase reflection bandwidth tends to be narrower. Even so, all simulated curves exhibit relatively flat profiles for different incident angles, indicating good in-phase reflection characteristics over the operating frequencies. All in all, our proposed AMC structure features light weight, miniaturized dimensions and wider in-phase reflection bandwidth, which shall be very promising for conformal array antenna designs.

2.3.1.3 Design and Analysis of the Conformal Subarray A

In our previous work, a conformal microstrip array was developed to achieve high gain and broadside radiation characteristics [20]. However, the microstrip antenna suffer from narrowband operation due to its intrinsic high Q-factor. In this work, a conformal dipole array backed by the miniaturized AMC (designed in Section II-B), i.e., the conformal Subarray A, is proposed to achieve high gain while maintaining ultrawide bandwidth. Instead of half-wavelength dipoles, the full-wavelength dipoles [21] are utilized to increase the gain of the conformal Subarray A due to their more directive radiation characteristics. However, considering their relatively high input impedances and steep frequency-dependent characteristics [22], we have made efforts to solve the issues associated with impedance matching and limited bandwidth. This part presents the specific details of designing a wideband high-gain conformal Subarray A and analyzing the influence of the balloon curvature on the radiation performance of the antenna.

Figure 2.48 depicts the structural geometry and planar expanded layout of the conformal Subarray A. As shown in Fig. 2.48a and d, the conformal dipole array is placed at the center of the AMC to achieve high gain and ultrawide bandwidth. To reduce the input impedance of the full-wavelength dipole and mitigate its drastic variation versus frequency, the dipole arms are widened at the ends to be a dumbbell shape. In this way, the impedance bandwidth can be broadened due to the increase in equivalent arm width. To further enhance the antenna gain, four full-wavelength dipoles are utilized and cross-fed in series via a central assembly line, which helps to sharpen the beam and reduce the sidelobe level. Utilizing more dipoles is basically beneficial to achieving higher gain and directivity, while leading to reduced antenna operating bandwidths due to deterioration in impedance matching and radiation patterns for such series-fed antenna array with frequency-dependent feeding networks and phase relationships. As depicted in Fig. 2.48b and c, a wideband balun is also designed for feeding the conformal subarray, which is implemented on a 1-mm-thick Rogers 5880 substrate. In Fig. 2.48c, the exponential tapered line can be described as

$$y(t) = \frac{W_b}{2} e^{-t}, 0 \leq t \leq L_{b2} \tag{2.5}$$

where W_b and L_{b2} are the line width and length, respectively. Figure 2.48d is a cross-sectional view of the structure, showing the dipole array, the AMC and the gap between them. In our design, the gap is replaced with a piece of 1.5-mm-thick foam, whose dielectric constant is approximately equal to air. Additionally, all substrate material in Fig. 2.48d is Rogers 5880 with relatively low dielectric loss of 0.0009.

Figure 2.49 shows the simulated reflection coefficients of the conformal Subarray A for different cylindrical radii. Benefiting from the positively coupling effects between the AMC and dipoles, the antenna operating bandwidth can be widened to be from 2.24 to 2.88 GHz for relatively larger values of the balloon radius. Moreover, it is clearly seen that the antenna operating bandwidth keeps nearly unchanged for the balloon radius R more than 300 mm, but decreasing remarkably for smaller radii due to poor impedance matching for the cases of severe bending.

Figure 2.50a depicts the simulated co-polarization and cross-polarization gains of the conformal Subarray A for different radii. It is clearly seen that the co-polarization

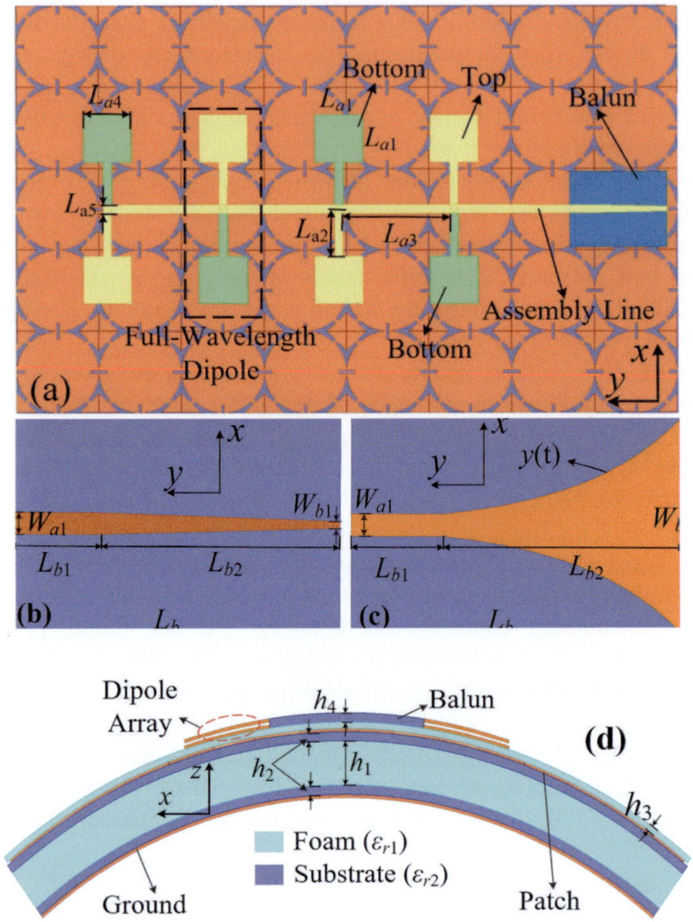

Fig. 2.48 **a** Top view of the planar expanded structure of the conformal Subarray A; **b** top and **c** bottom layers of wideband balun; **d** cross-sectional view of the conformal Subarray A. Size parameters: $L_{a1} = 17.375$, $L_{a2} = 17.375$, $L_{a3} = 46.3$, $L_{a4} = 9.2$, $L_{a5} = 1.9$, $L_{\mathrm{b}} = 35$, $W_{\mathrm{b}} = 28$, $L_{b1} = 7$, $L_{b2} = 28$, $W_{\mathrm{b}1} = 1.8$, $\varepsilon_{r1} = 1$, $\varepsilon_{r1} = 2.2$, $h_1 = 0.25$, $h_2 = 2.5$, $h_3 = 1.5$, $h_4 = 1$ (the units are mm except for ε_{r1} and ε_{r2})

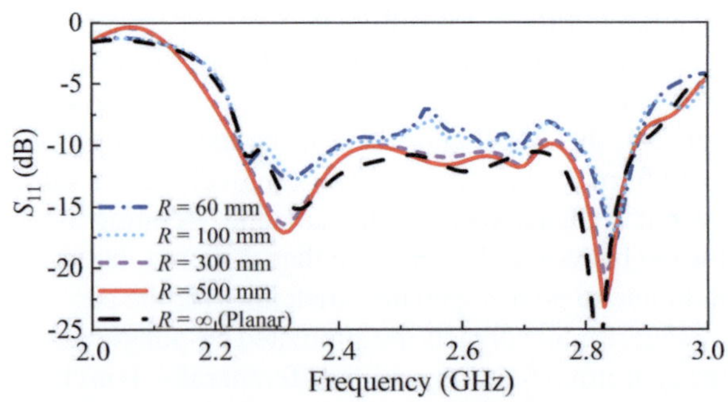

Fig. 2.49 Simulated reflection coefficients of the conformal Subarray A for different radii

gain decreases with increasing the balloon curvature due to the reduced effective radiating aperture. Although the cross-polarization level rises with decreasing *R*, which always keeps about 10 dB lower than the cross-polarization gain in the operating band, thus having little influence on the radiation performance of the antenna. Figure 2.50b and c depict the simulated radiation patterns for different bending radii in the *xz* and *yz* planes, respectively. Obviously, the radiation patterns feature wider beamwidths along the bending direction (i.e., the *x* direction) for larger balloon curvatures, while maintaining the directional shapes in the *yz* plane. Obviously, less bending leads to larger effective radiating aperture and thus a shaper beam and higher gain. Table 2.8 compares the radiation performance of the conformal Subarray A for different cylindrical radii, including the co-polarization and cross-polarization gains, and half-power beam width (HPBW) in both the *xz* and *yz* planes. Clearly, the conformal Subarray A exhibits good radiation performance for the balloon radius more than 300 mm.

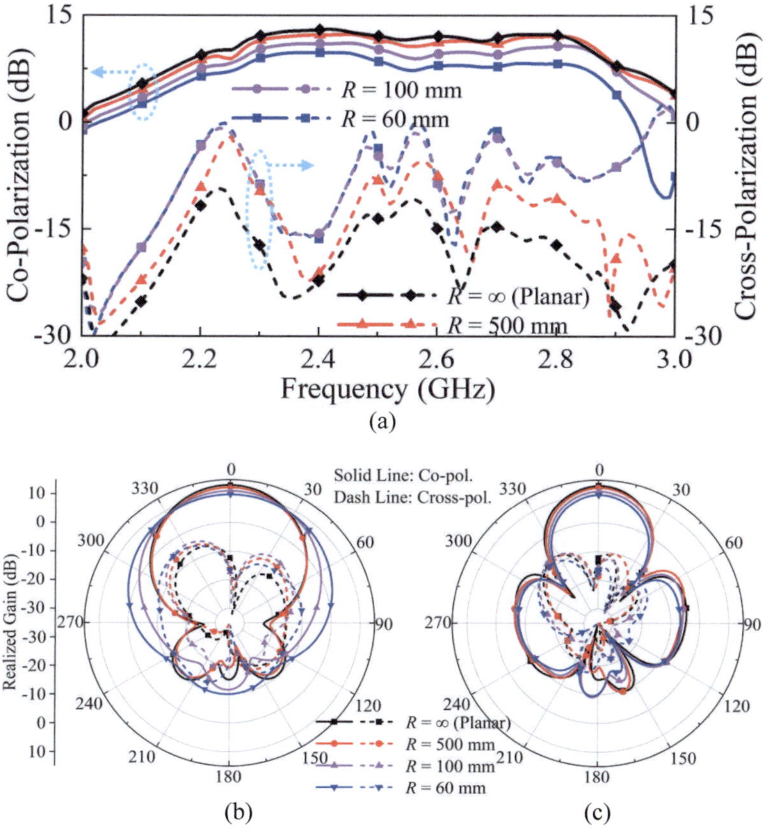

Fig. 2.50 Simulated results of the conformal Subarray A for different radii: **a** co-polarization and cross-polarization gains versus frequency; **b** and **c** radiation patterns in the *xz* and *yz* planes, respectively

Table 2.8 Simulated performance of the conformal Subarray A for different cylindrical radii at the central frequency

Radius (mm)	Co-Pol. Gain (dB)	Cross-Po.l Gain (dB)	HPBW (°)	
			xz plane	*yz* plane
60	9.8	– 1.6	90.6	33.3
100	11.05	– 3.9	66.8	33.9
300	12.23	– 4.5	49.71	34.1
400	12.29	– 4.8	48.43	33.7
500	12.30	– 5.2	47.86	34.1
600	12.43	– 5.3	47.13	32.8
Planar (∞)	12.91	– 7.8	45.41	33.1

2.3.1.4 Design and Analysis of the Conformal Subarray B

Benefitting from their wideband radiation characteristics, typical endfire radiators like log-periodic antennas have been developed for conformal systems [23]. Such log-periodic structures generally require a number of elements for achieving the log-periodic frequency-dependent property for broadband operation. For the log-periodic antenna, the embarrassed feeding position at the forefront may cause partial beam blocking, or a special feeding balun from the backend to forefront is required which imposes the design complexity. To avoid such feeding issue, this part presents the specific details of designing an innovative conformal endfire antenna using size-tapered dipoles, i.e., the conformal Subarray B, achieving ultrawide operating bandwidth while featuring a relatively compact geometry and simple feeding structure. Moreover, by taking full advantage of the room of the shared aperture, the size of radiating aperture along the circumferential direction is enlarged by utilizing two pairs of serial-fed dipoles in parallel, thus increasing the gain and achieving the required fan-shaped beam for balloon-based communication applications. Detailed simulation analyses are carried out to examine the influences of structural bending and conforming on the radiation performance of the conformal Subarray B.

Figure 2.51 shows the structural geometry of the conformal Subarray B, which includes 2 × 3 tapered dumbbell-shaped half-wavelength dipoles that are fed in series and then parallel via a power divider. The expanded layout of the conformal two-way T-junction divider is depicted in Fig. 2.51b, for which the patterns on the top and bottom layers are exactly the same. Quarter-wavelength transformers are used to guarantee good impedance matching for the divider. Figure 2.51c shows the planar expanded structure of the conformal tapered dipole array, where three size-tapered dipole elements are separate with optimized spacing to achieve the desired radiation characteristics. By sharing the radiating aperture, another group of 1 × 3 size-tapered dipoles are utilized to not only increase the gain but also producing the desired fan-shaped endfire beam as shown in Fig. 2.42.

Figure 2.52 illustrates the simulated reflection coefficients of the conformal Subarray B for different cylindrical radii. Clearly, the proposed conformal Subarray

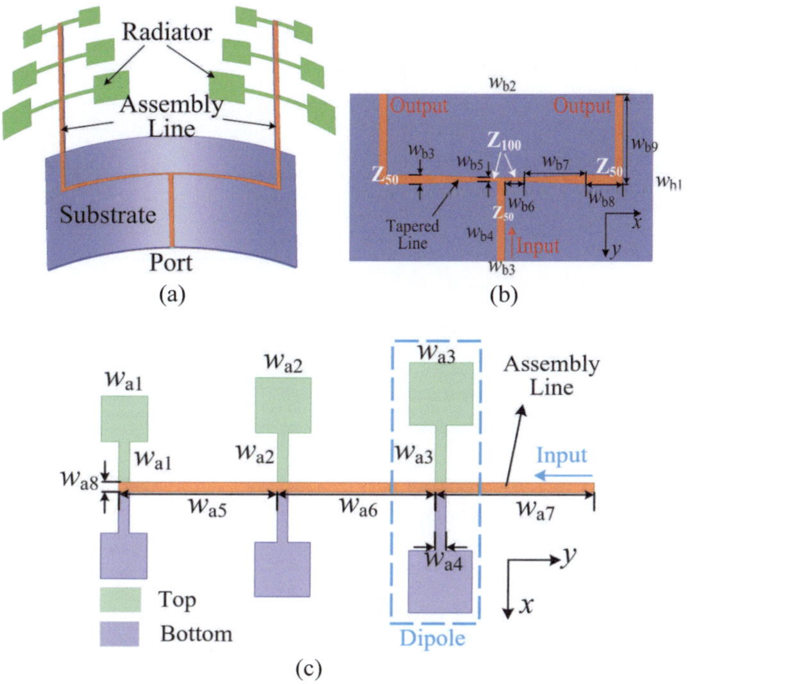

Fig. 2.51 **a** Perspective view of the conformal Subarray B; **b** and **c** top views of the planar expanded structures of the power divider and the tapered dipole array. Size parameters: $w_{a1} = 36$, $w_{a2} = 48$, $w_{a3} = 60$, $w_{a4} = 2$, $w_{a5} = 20$, $w_{a6} = 20$, $w_{a7} = 30$, $w_{a8} = 1.9$, $w_{b1} = 60$, $w_{b2} = 100$, $w_{b3} = 1.8$, $w_{b4} = 30$, $w_{b5} = 1$, $w_{b6} = 5$, $w_{b7} = 30$, $w_{b8} = 5$, $w_{b9} = 21.4$ (the units are mm)

B (for R more than 300 mm) and its planar version exhibit almost identical impedance bandwidth of 2.08–3.03 GHz (or relative bandwidth of about 39.6%). As the balloon radius decreases to be less than 100 mm, impedance matching is severely destroyed resulting in a reduced operating bandwidth of the antenna, which is still above 29.6% (or 2.18–2.89 GHz) for R equal to 60 mm. Obviously, utilizing the size-tapered dipole elements introduces new resonant frequencies, thus widening the impedance bandwidth.

Figure 2.53 illustrates the simulated co-polarization and cross-polarization gains of the conformal Subarray B for different radii. It is seen that the bending causes performance degradation in antenna gain, i.e., the variation of the co-polarization and cross-polarization levels. It is seen that for smaller values of the balloon radius, the co-polarization gain decreases remarkably and the cross-polarization level increases significantly. This phenomenon is also clearly illustrated by Fig. 2.53b and c, which depict the simulated radiation patterns for different bending radii in the xy and yz planes, respectively. It is seen that the increase of balloon curvature plays very minor role in reshaping the radiation patterns, while leading to considerably increased cross-polarization levels in the xy plane. The result can be readily understood

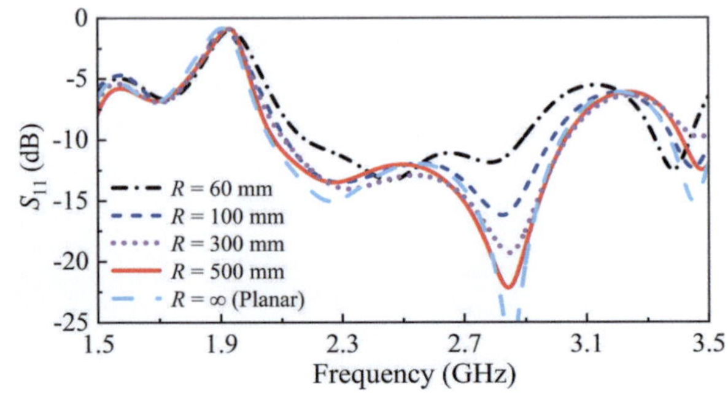

Fig. 2.52 Simulated reflection coefficients of the conformal Subarray B for different cylindrical radii

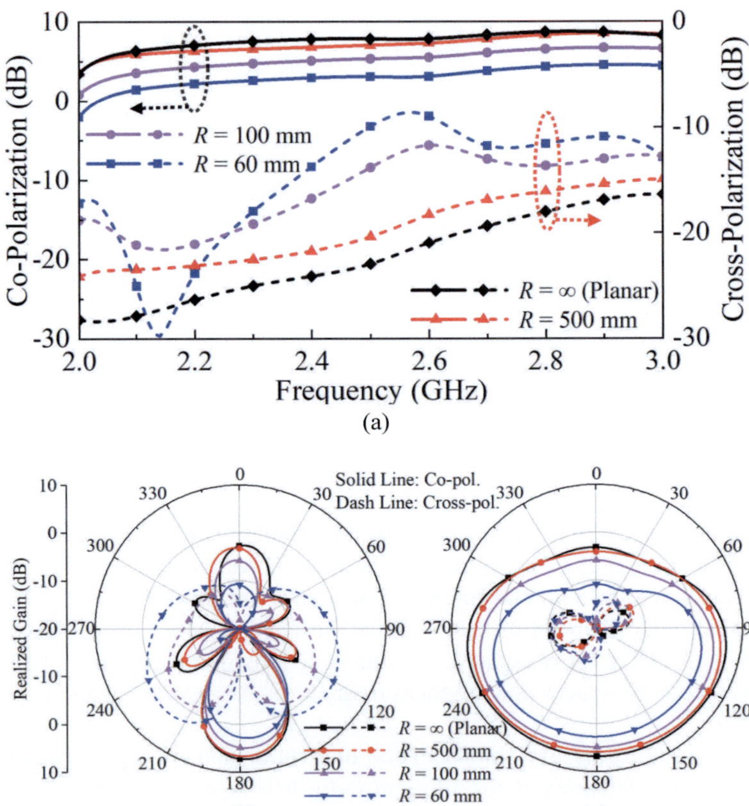

Fig. 2.53 Simulated results of the conformal Subarray B for different radii: **a** co-polarization and cross-polarization gains versus frequency; **b** and **c** radiation patterns in the *xy* and *yz* planes, respectively

Table 2.9 Simulated performance of the conformal Subarray B for different cylindrical radii at the central frequency

Radius (mm)	Co-Pol. Gain (dB)	Cross-Po.l Gain (dB)	1.5-dB beam width	
			xy plane	*yz* plane
50	1.27	3.7	–	–
60	2.9	2.4		–
80	4.4	0.5	40.3	166
100	5.1	– 1.2	32.83	180.2
300	6.72	– 9.9	26.69	175.1
400	6.75	– 12	26.64	174.5
500	6.79	– 13.7	26.45	175.6
600	6.83	– 14.8	26.21	175.5
Planar (∞)	7.20	– 18.3	25.49	176.1

from the structural geometry as shown in Fig. 2.51a. The conformal Subarray B is comprised of two endfire radiating elements, the polarization relationship between which becomes quite complicated for larger balloon curvatures, thus resulting in higher cross-polarization levels in the combined radiation patterns. As shown in Table 2.9, for the balloon radius more than 300 mm, the conformal Subarray B achieves a relatively narrow beamwidth in the xy plane but much wider beamwidth in the yz plane, which is critical for contributing to pattern combination for half-space beam coverage meanwhile achieving a moderate gain.

2.3.2 *Measurement, Simulation and Analysis*

2.3.2.1 Measurement and Simulation of the Conformal Subarray A

To reduce the cost, only one Subarray A and one Subarray B were fabricated. However, this is enough to validate the designs and demonstrate the radiating performance of whole conformal heterogeneous array considering its beam switching characteristics. Figure 2.54 shows the fabricated prototypes of the conformal subarrays A and B, as well as the experimental setup in an anechoic chamber. During prototype manufacturing, all metalized parts are copper with a thickness of 17 μm, except for all the dipoles utilizing H_{62}-type copper patches with thickness of 100 μm. As depicted in Fig. 2.54a and c, those copper patches are cut and shaped as designed patterns, and then soldered to the central assembly line. The radio frequency signals are fed through the SMA at the ends of the printed circuit boards. As shown in Fig. 2.54a–c, the fabricated Subarrays A and B have expanded planar dimensions of $2.112\lambda_0 \times 1.32\lambda_0 \times 0.044\lambda_0$ and $1.224\lambda_0 \times 1.088\lambda_0 \times 0.008\lambda_0$, respectively. Based on the simulation analysis results as discussed in section II-C and II-D, the balloon radius is required to be more than 300 mm for guaranteeing the radiation performance of the conformal antenna operating at the frequency band. Moreover, considering larger balloon radius is beneficial to offering stronger floating capability in practical communication applications, a cylindrical balloon made of polyvinyl chloride with a radius of 500 mm was manufactured. As depicted in Fig. 2.54d, benefitting from their ultrathin profiles and excellent flexibility, the conformal radiators are easily attached with the inflatable balloon. For the convenience of prototype measurement, the antenna was attached to the balloon by using the transparent adhesive tapes. The tapes are ultrathin thus not causing any negative effect on the antenna radiation performance. Considering the conformal Subarray A is a broadside radiator, its radiation performance can be measured directly using the experimental setup as shown in Fig. 2.54d. In comparison, the conformal Subarray B is an endfire radiator with main beam downwards, whose radiating performance was measured by rotating the balloon by 90° to align the beam with that of a standard gain horn.

Figure 2.55 shows the measured and simulated results of reflection coefficient and realized gain for the conformal Subarray A. As depicted in the Fig. 2.55a, the measured and simulated S_{11} curves have the similar shape, although there is a minor frequency offset which may be caused by dielectric constant or manufacturing errors. The measured results indicate that the operating bandwidth for S_{11} < –– 0 dB is from 2.1 to 2.82 GHz (720 MHz), which is slightly wider than the simulated one of 2.24 –2.88 GHz (640 MHz). The widened bandwidth mainly results from various losses in experiments, such as extra losses induced by soldering process and nonideal dielectric material. Figure 2.55b depicts the measured and simulated realized gain versus the operating frequency. The simulated and measured curves also have very

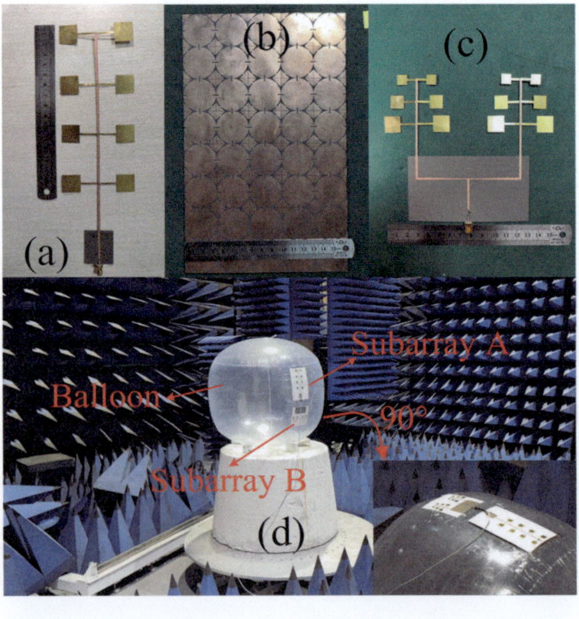

Fig. 2.54 Photographs of the **a** full-wavelength dipole array, **b** flexible AMC reflector of Subarray A, **c** Subarray B and **d** experimental setup in an anechoic chamber

similar shape, and there exists a minor frequency offset. The measured gain over the operating band is slightly lower than the simulated one, which is caused by material and fabrication losses during experiment. The measured results show that the gain is 11.45 dB at the central frequency, achieving its peak value of around 12 dB at 2.3 GHz. Good consistency between the measured and simulated results has confirmed the effectiveness of our design.

Figure 2.56a–f shows the measured and simulated radiation patterns of the conformal Subarray A in the *xz* and *yz* planes for different frequencies, respectively. It is seen that the simulated and measured results are in good agreement at the central frequency and lower band, while the measured gain at high frequency is slightly lower than the simulated one. This is due to the frequency offset toward lower frequency band for the manufactured prototype, as depicted in Fig. 2.55a. Additionally, the gain at the boundary of higher operating band drops quicker than that for the lower band, resulting in a slightly lower measured gain than simulation at 2.88 GHz. Although the antenna radiation performance is slightly degraded at high band, the main beam direction does not change and would have little impact on practical utilization. At the central frequency, the measured HPBW of the *xz* and *yz* planes are 48° and 36°, respectively.

Fig. 2.55 Measured and Simulated results for the conformal Subarray A: **a** reflection coefficient and **b** realized gain

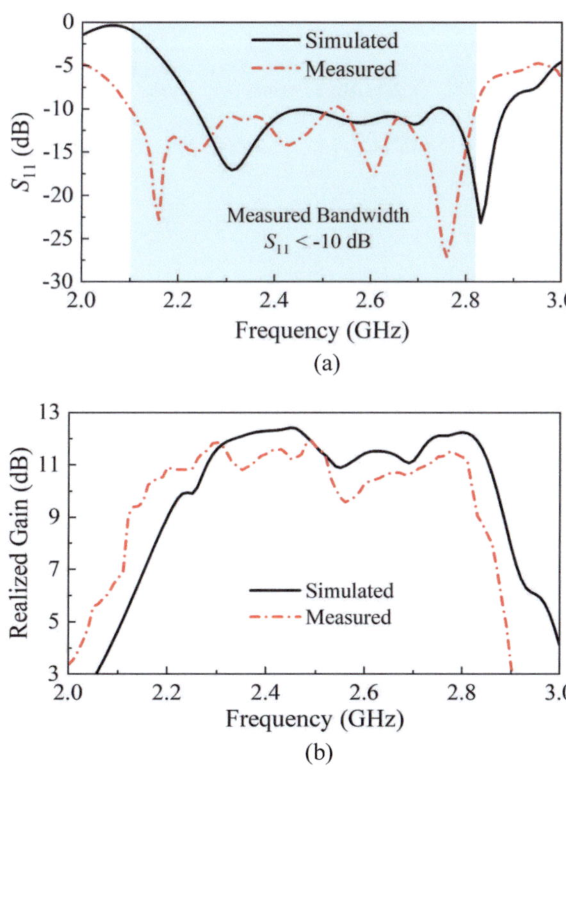

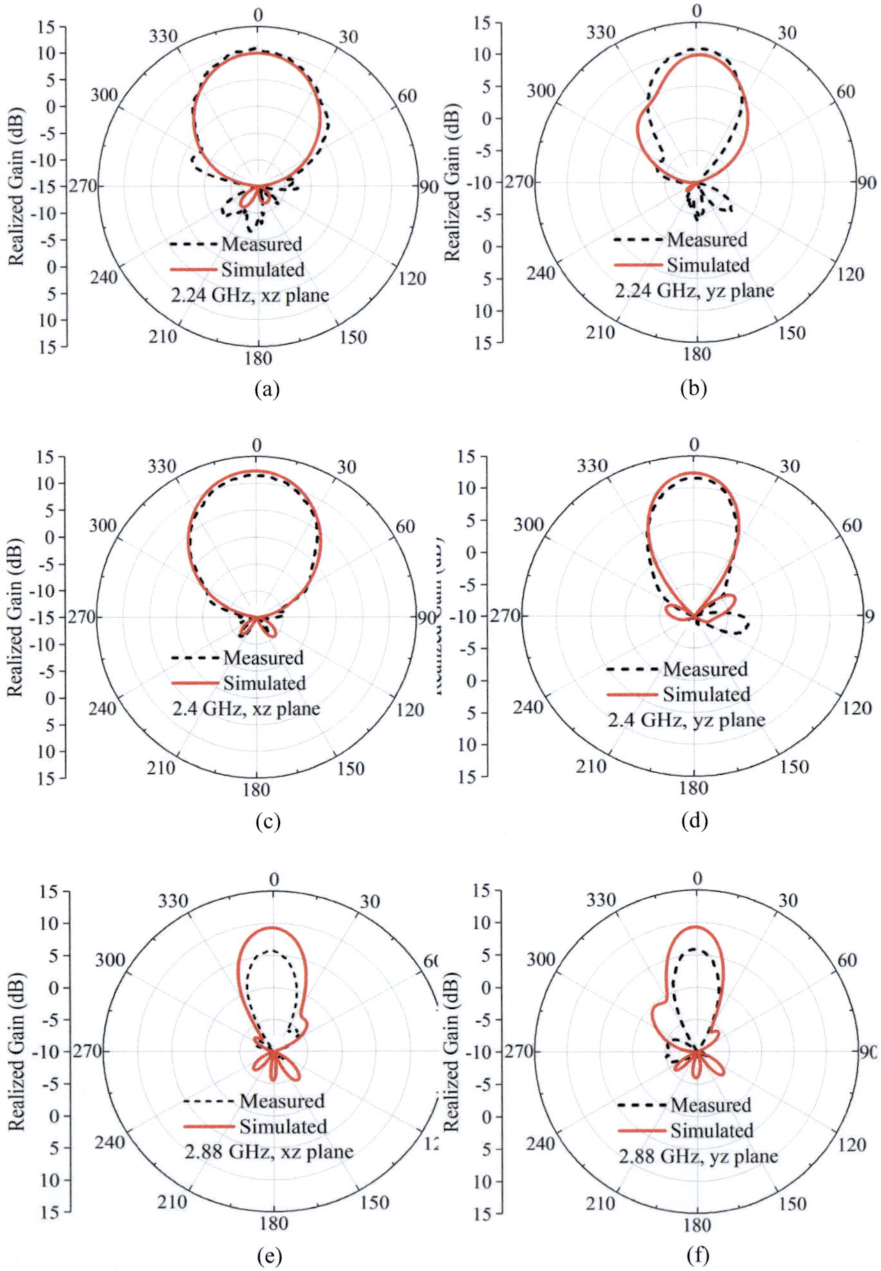

Fig. 2.56 Measured and simulated radiation patterns of the conformal Subarray A at different frequencies and different principal planes: **a** *xz* and **b** *yz* planes at 2.24 GHz; **c** *xz* and **d** *yz* planes at 2.4 GHz; **e** *xz* and **f** *yz* planes at 2.88 GHz

2.3.2.2 Measurement and Simulation of the Conformal Subarray B

Figure 2.57 shows the simulated and measured results of reflection coefficient and realized gain for the conformal Subarray B. As depicted in Fig. 2.57a, the measured results agree very well with simulation. The measured results show that the operating bandwidth for $S_{11} < -10$ dB is from 2.08 to 3.12 GHz (1.04 GHz), which is slightly wider than the simulated bandwidth of 2.08–3.03 GHz (0.95 GHz). The widened bandwidth is mainly benefited from extra losses during experiment. Figure 2.57b illustrates the measured and simulated realized gain versus the operating frequency. Although the measured gain is slightly lower than the simulation due to the additional losses introduced in the measurements, the trend of the measured curve is almost the same as that of simulated one. Additionally, the measured results show that the gain is 6.37 dB at the central frequency and the peak gain is around 7.7 dB at 2.9 GHz.

Figure 2.58a–f shows the simulated and measured radiation patterns for the conformal Subarray B in the *xz* and *yz* planes for different frequencies, respectively. It can be found that the simulated and measured results are in good agreement in the *xy* plane, while the measured beam widths are slightly narrower than the simulated results in the *yz* plane. This minor discrepancy can be attributed to alignment errors during measurement. At the central frequency, the measured HPBW are around 36° and 171° in the *xy* and *yz* planes, respectively.

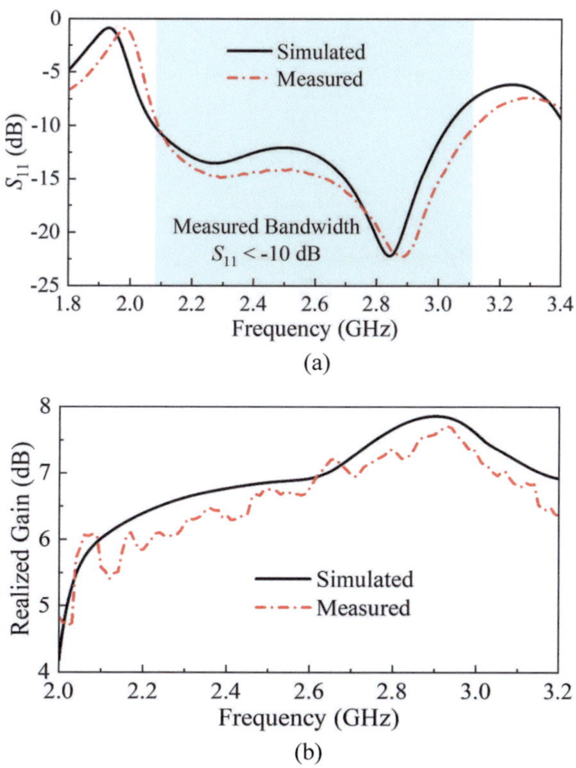

Fig. 2.57 Measured and simulated results for the conformal Subarray B: **a** reflection coefficient and **b** realized gain

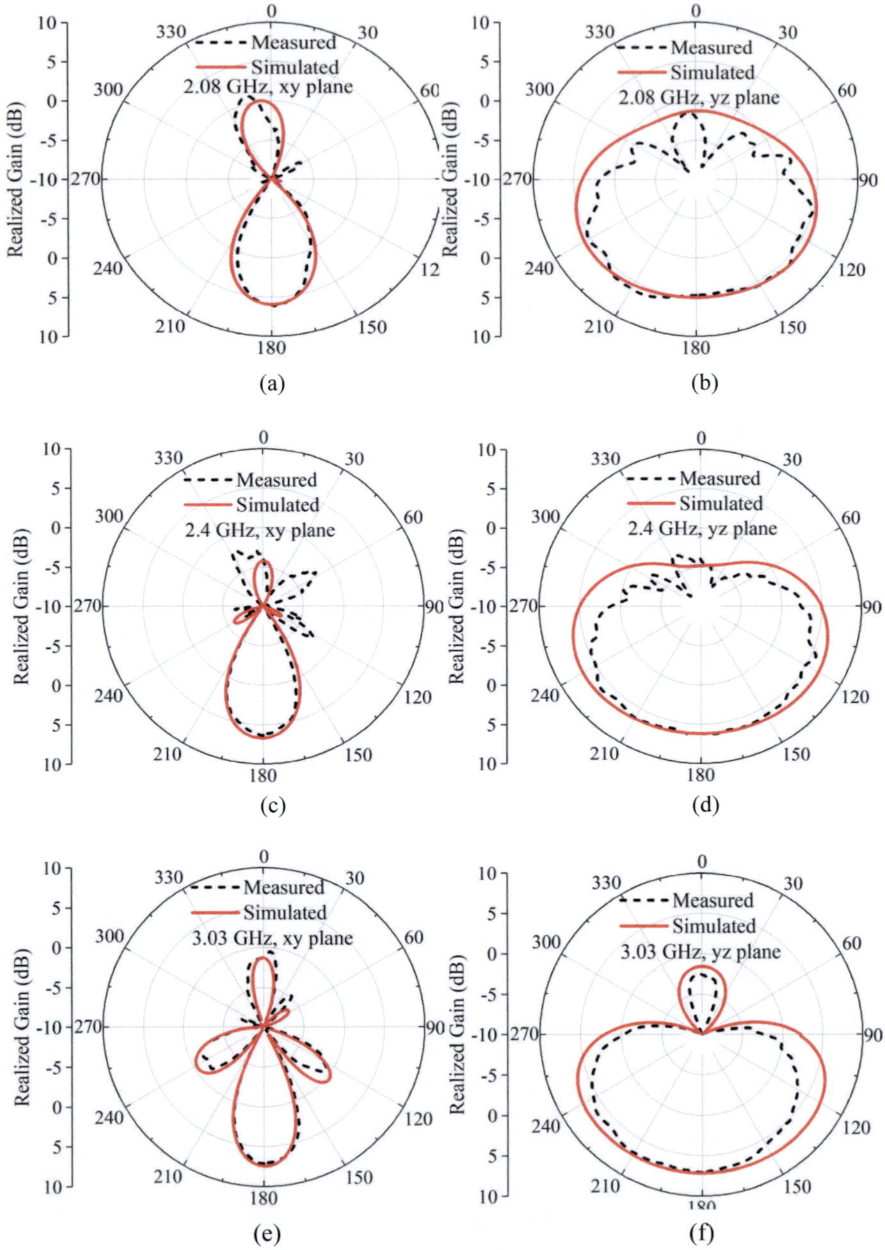

Fig. 2.58 Measured and simulated radiation patterns of the conformal Subarray B at different frequencies and different principal planes: **a** *xy* and **b** *zy* planes at 2.08 GHz; **c** *xy* and **d** *zy* planes at 2.4 GHz; **e** *xy* and **f** *zy* planes at 3.03 GHz

2.3.2.3 Analysis of Conformal Heterogeneous Array

The conformal Subarrays A and B are designed to have similar widths, thus very beneficial to form a shared radiating aperture. Utilizing nine conformal subarray pairs, a balloon-based conformal heterogeneous array antenna can then be implemented. The measured bandwidth of the Subarray A is 2.1–2.82 GHz, which lies well within the bandwidth of the Subarray B (i.e., 2.08–3.12 GHz), thus leading to an ultrawide operating bandwidth of the conformal heterogeneous array. Although the gain of the Subarray B is lower than that of the Subarray A, it does not cause negative impact on practical utilization. This is because the high-gain Subarray A is used for long-range air-to-ground or air-to-air communications, whereas the moderate-gain Subarray B is used for short-range air-to-ground communications requiring large angular coverage. By combining the radiation patterns of those two conformal arrays, a half-space beam coverage can be well achieved, as illustrated in Fig. 2.42. Considering the beam widths of the Subarrays A and B in the azimuthal plane, a 360° beam coverage is achieved by using nine conformal subarray pairs spaced uniformly around the cylindrical balloon surface. In the elevational plane, the radiation pattern of the Subarray B exhibits a beam width wider than 170°, the part of which overlaps well with that of the Subarray A (with beam width of 36°). Therefore, the conformal heterogeneous array can achieve a wide beam coverage of 360° × 108°, which is very beneficial for emergency communication scenarios in the mountainous and dense-forest environment.

References

1. R. L. Carrel, *Analysis and Design of the Log-Periodic Dipole Antenna.* Urbana, Illinois, USA: University of Illinois, 1961.
2. C. A. Balanis, *Antenna Theory: Analysis and Design.* New York: Wiley, 1992.
3. C. Ding, X. Gao, K. Yang, J. Song, R. Li, X. Bu, and J. An. "Comparative studies on multi-carrier transmission schemes in mountainous and dense forest environment," *Sci. Rep.*, vol. 12, no. 1, p. 16782, Oct. 2022.
4. L. Gupta, R. Jain, and G. Vaszkun, "Survey of important issues in UAV communication networks," *IEEE Commun. Surv. Tut.*, vol. 18, no. 2, pp. 1123–1152, 2nd Quart., Nov. 2015.
5. M. Mozaffari, W. Saad, M. Bennis, Y.-H. Nam, and M. Debbah, "A tutorial on UAVs for wireless networks: applications, challenges, and open problems," *IEEE Commun. Surveys Tuts.*, vol. 21, no. 3, pp. 2334–2360, Mar. 2019.
6. Y. Zeng, R. Zhang, and T. J. Lim, "Wireless communications with unmanned aerial vehicles: opportunities and challenges," *IEEE Commun. Mag.*, vol. 54, no. 5, pp. 36–42, May. 2016.
7. S. H. Alsamhi, F. A. Almalki, F. Afghah, A. Hawbani, A. V. Shvetsov, B. Lee, and H. Song, "Drones' edge intelligence over smart environments in B5G: blockchain and federated learning synergy," *IEEE Trans. Green Commun. Netw.*, vol. 6, no. 1, pp. 295–312, Mar. 2022.
8. S. H. Alsamhi, M. S. Ansari, O. Ma, F. Almalki, and S. K. Gupta, "Tethered balloon technology in design solutions for rescue and relief team emergency communication services," *Disaster Med. Publ. Health Preparedness*, vol. 13, no. 2, pp. 203–210, May. 2018.
9. S. H. Alsamhi, F. A. Almalki, O. Ma, M. S. Ansari, and M. C. Angelides, "Performance optimization of tethered balloon technology for public safety and emergency communications," *Telecommun. Syst.*, vol. 75, no. 2, pp. 235–244, May. 2019.

10. S. H. Alsamhi, M. S. Ansari, and N. S. Rajput, "Disaster coverage predication for the emerging tethered balloon technology: capability for preparedness, detection, mitigation, and response," *Disaster Med. Publ. Health Preparedness*, vol. 12, no. 2, pp. 222–231, Apr. 2018.
11. S. A. Khaleefa, S. H. Alsamhi, and N. S. Rajput, "Tethered balloon technology for telecommunication, coverage and path loss," in Proc. *IEEE Students Conf. Electr. Electron. Comput. Sci.*, 2014, pp. 1–4.
12. E. Chang, S. A. Long, and W. F. Richards. "An experimental investigation of electrically thick rectangular microstrip antennas," *IEEE Trans. Antennas Propag.*, vol. 34, no. 6, pp. 767–772, Jun. 1986.
13. K. Klionovski and A. Shamim, "Physically connected stacked patch antenna design with 100% bandwidth," *IEEE Antennas Wireless Propag. Lett.*, vol. 16, pp. 3208–3211, Oct. 2017.
14. Y. Fang, L. S. Wu, L. F. Qiu and Y. P. Zhang, "A method of introducing coupling null by shorting pins for stacked microstrip patch antenna array," *IEEE Trans. Antennas Propag.*, vol. 70, no. 7, pp. 6030–6035, Jul. 2022.
15. L. Josefsson and P. Persson, *Conformal Array Antenna Theory and Design.* Hoboken, NJ, USA: Wiley, 2006.
16. H. Malekpoor and S. Jam, "Improved radiation performance of low profile printed slot antenna using wideband planar AMC surface," *IEEE Trans. Antennas Propag.*, vol. 64, no. 11, pp. 4626–4638, Nov. 2016.
17. A. Foroozesh and L. Shafai, "Investigation into the application of artificial magnetic conductors to bandwidth broadening, gain enhancement and beam shaping of low profile and conventional monopole antennas," *IEEE Trans. Antennas Propag.*, vol. 59, no. 1, pp. 4–20, Jan. 2011.
18. D. Sievenpiper, L. Zhang, R. F. J. Broas, N. G. Alexopolous, and E. Yablonovitch, "High-impedance electromagnetic surfaces with a forbidden frequency band," *IEEE Trans. Microw. Theory Techn.*, vol. 47, no. 11, pp. 2059–2074, Aug. 1999.
19. D. Chen, W. Yang, and W. Che, "High-gain patch antenna based on cylindrically projected EBG planes," *IEEE Antennas Wireless Propag. Lett.*, vol. 17, no. 12, pp. 2374–2378, Dec. 2018.
20. C. Ding, X. Gao, W. Tang, H. Liu, C. Zhu, X. Bu, and J. An, "A Balloon Conformal Array Antenna of Broadened Bandwidth, High Gain and Steerable Directional Beam Fed by A Flexible Phase Shifting Network," *IEEE Trans. Antennas Propag.*, vol. 71, no.11, pp. 8525–8536, Nov. 2023.
21. C. Ding, H. Sun, H. Zhu and Y. Jay Guo, "Achieving wider bandwidth with full-wavelength dipoles for 5G base stations" *IEEE Trans. Antennas Propag.*, vol. 68, no. 2, pp. 1119–1127, Feb. 2020.
22. C. Ding, B. Jones, Y. J. Guo and P. Qin, "Wideband matching of full-wavelength dipole with reflector for base station," *IEEE Trans. Antennas Propag.*, vol. 65, no. 10, pp. 5571-5576, Oct. 2017.
23. X. Gao, Z. Shen, and C. Hua, "Conformal VHF log-periodic balloon antenna," *IEEE Trans. Antennas Propag.*, vol. 63, no. 6, pp. 2756–2761, Jun. 2015.
24. R. A. Burberry, *VHF and UHF Antennas.* London: Peter Peregrinus Ltd., 1997.

Chapter 3
Conformal Array Antennas for Mobile Platforms

The rapid development of unmanned systems and wearable technology has propelled mobile-platform conformal array antennas to the forefront of wireless communication research. These arrays are distinguished by their minimal disturbance to the aerodynamic profile of the host platform, exceptional space utilization efficiency, and proactive robustness in complex electromagnetic environments. Traditional planar array antennas face significant bottlenecks on mobile platforms, including limited scanning ranges, inadequate polarization flexibility, and poor integration with the platform structure. Conformal array antennas overcome these constraints by conforming to the host surface, significantly improving beam coverage, enhancing attitude compatibility, and ultimately delivering superior communication performance for mobile scenarios. This chapter focuses on conformal array antennas for mobile platforms. Addressing critical emergency response scenarios—such as firefighting and military operations—we confront the core challenges of dynamic environmental adaptability and multifunctional integration. To this end, we present innovative solutions: a cylindrical conformal array antenna integrated onto unmanned aerial vehicles (UAVs) and a spherical conformal array antenna integrated into helmets for typical emergency rescue scenario as shown in Fig. 3.1.

The proposed two mobile-platform conformal array antenna demonstrate hierarchical advancements in reconfigurable design methodology, bandwidth, and coverage. The UAV-oriented conformal array antenna comprised of three conformal subarrays, where each subarray includes nine quad-polarization reconfigurable radiating elements. The presented conformal array exhibits a 3-dB axial-ratio bandwidth of up to 1.3% at around 1.5 GHz and − 10-dB impedance bandwidth of 7.7% for two circular-polarization modes and 7.3% for two linear-polarization modes, respectively. Besides, the realized gain is over 14 dBi for all reconfigurable polarization modes over a relatively wide beam coverage range of $50° \times 180°$. This work addresses the challenge of communication instability in UAV systems induced by payload-imposed constraints and orientation alterations within topographically complex

X. Gao et al., *Conformal Array Antennas and Beamforming Technologies for Emergency Communication Applications*, Modern Antenna,
https://doi.org/10.1007/978-981-95-2905-6_3

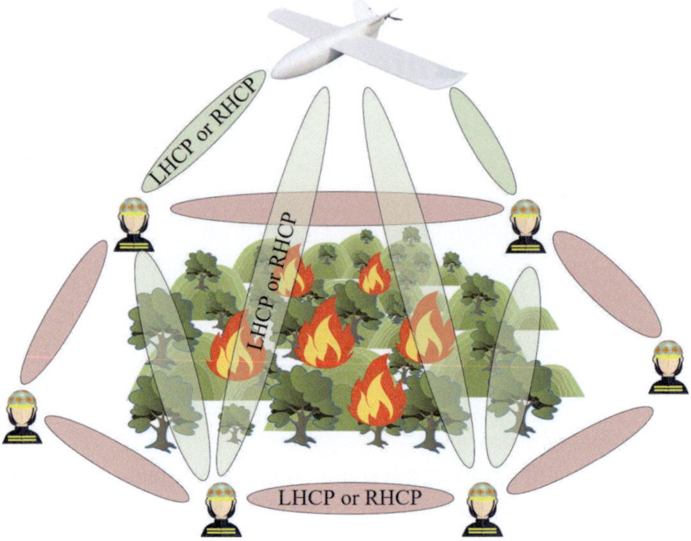

Fig. 3.1 Schematic showing the proposed drone-based quad-polarization reconfigurable conformal array antenna and helmet-based hemispherical-conformal wearable array antenna empowered Wireless body area network (WBAN) applications in typical emergency rescue scenarios with forest fires

environments, while confirming the technical viability of deploying polarization-reconfigurable phased array antennas on mobile aerial platforms. Compared to UAV-oriented conformal array antenna, the helmet-based hemispherical-conformal array antenna innovatively employs a crossed-dipole coupled with parasitic patch structure, achieving a 29.6% axial ratio bandwidth and 69.8% impedance bandwidth while supporting LHCP/RHCP dual-mode reconfiguration with PIN diodes. Its ultra-wide beam coverage of $110° \times 360°$ significantly enhances communication robustness between ground and aerial terminals, while the simplified feeding topology with PIN diodes meets the lightweight and low-profile requirements for helmet-mounted systems. The two research endeavors are tightly interlinked, enabling a transition from single cylindrical conformal arrays to two-dimensional spherical conformal arrays under polarization-reconfigurable conditions. Performance-wise, both bandwidth and coverage range have demonstrated marked improvements. Consequently, the work presented in this chapter advances mobile-platform conformal array antenna design, enabling highly reliable communication links for scenarios such as emergency rescue and disaster monitoring.

3.1 A Quad-Polarization Reconfigurable Conformal Array Antenna for Unmanned Aerial Vehicle Communication Applications

In this section, a quad-polarization reconfigurable conformal array of wide coverage range and high gain is presented for UAV communication systems. The proposed single-port excited miniaturized element utilizes a low-complicity reconfigurable feeding network to realize two orthogonal LPs, LHCP, and RHCP polarization modes. A quad-polarization reconfigurable UAV conformal sparse array is proposed, designed, and characterized. The measurement results show that the 3 dB AR bandwidth is around 1.3% for the CP mode, and the − 10 dB impedance bandwidths are 7.7 and 7.3% for the CP and LP modes. The measured gain of the conformal array is over 14 dBi for all polarization modes. A relatively wide beam coverage range of 50° × 180° is achieved. In summary, the good radiating performance of the proposed conformal sparse array antenna makes it potentially be applied to the UAVs with payload constraints for emergency communication applications.

3.1.1 Antenna Design and Analysis

3.1.1.1 Design of Polarization Reconfigurable Radiating Element

Figure 1 illustrates the geometrical configuration of the proposed radiating element that can operate in four reconfigurable polarization modes, including the − 45° and + 45° linear polarizations (LPs) as well as the left-hand (LH) and right-hand (RH) circular polarizations (CPs). The radiating element has a total dimension of 97 mm × 99 mm, which consists of two major parts, i.e., a dual-feed miniaturized microstrip patch and a small low-complicity reconfigurable feeding network. As shown in Fig. 3.2, those two parts are stacked together in such way that they share the same ground plane, which leads to a total thickness of $0.029\lambda_0$ for the whole structure.

As depicted in Fig. 3.2a, a miniaturized circular patch radiator with radius of r_{p1} is designed on a h_1-thick F_4BM220 substrate with relatively low dielectric constant of 2.2, where its ground plane is printed on the back side of the same substrate. For the proposed radiator, four rectangular slots with sizes of $l_{r1} \times w_{r1}$ are cut symmetrically to prolong the electric current path for size reduction. To realize quad-polarization reconfiguration and increase the axial-ratio (AR) bandwidth for CP modes, two feeding pins are utilized to excite the antenna with desired amplitude and phase relationship as determined by the polarization reconfigurable network. For whichever polarization mode, the total radiation field is the vector superposition of two orthogonal fields generated independently by the ports P1 and P2 as shown in Fig. 3.2a. Since the circular patch radiator operates in the TM_{11} mode for both orthogonal fields, the electric field strength is zero at the patch center. Hence, a

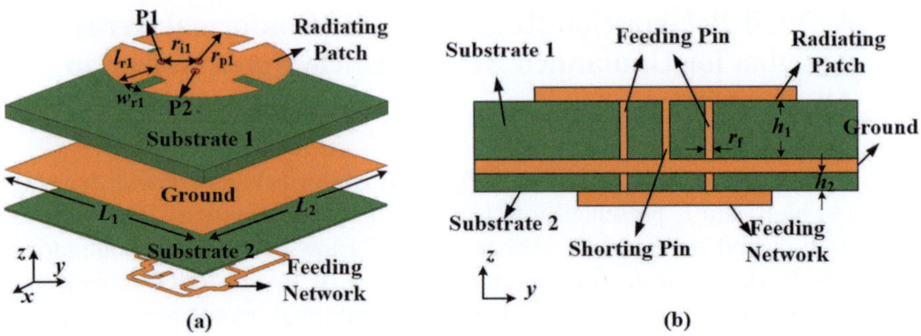

Fig. 3.2 Geometrical configuration of the polarization reconfigurable radiating element: **a** 3D-view; **b** side view. ($l_{r1} = 14$ mm $w_{r1} = 6$ mm, $r_{p1} = 34$ mm, $r_{i1} = 9$ mm, $L_1 = 97$ mm, $L_2 = 99$ mm, $h_1 = 5$ mm $h_2 = 0.8$ mm, $r_f = 0.6$ mm)

shorting pin located as shown in Fig. 3.2b has very little impact on the radiation field for all reconfigurable polarizations. However, such shorting pin can considerably simplify the direct-current (DC) biasing circuit design.

To analyze the performance of the radiating element, the software ANSYS High Frequency Structure Simulator (HFSS) is utilized for numerical simulation. Figure 3.3 shows the simulated scattering parameters of the radiating element when excited independently by the port 1. Clearly, the patch antenna has a − 10-dB impedance bandwidth of around 40 MHz at the frequency of 1.5 GHz, and a port isolation higher than 30 dB over the operation band. Besides, it can be found that the shorting pin has almost no impact on the S_{11} of the antenna, while the loading of four slots effectively reduces the radiator size by 10%. Figure 3.4a and b show the simulated current vector and electric field magnitude at 1.5 GHz, respectively. Figure 3.4 clearly shows the typical TM_{11}-mode distribution on the radiating element when excited by the port 1, where the electric field strengths at the port 2 and the element center are almost zero.

Fig. 3.3 Simulated scattering parameters of the radiating element when excited independently by the port 1 (the port 2 is terminated with a 50-Ω matched load)

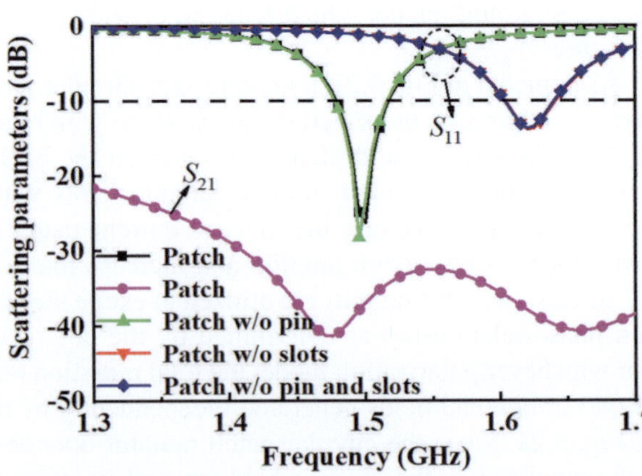

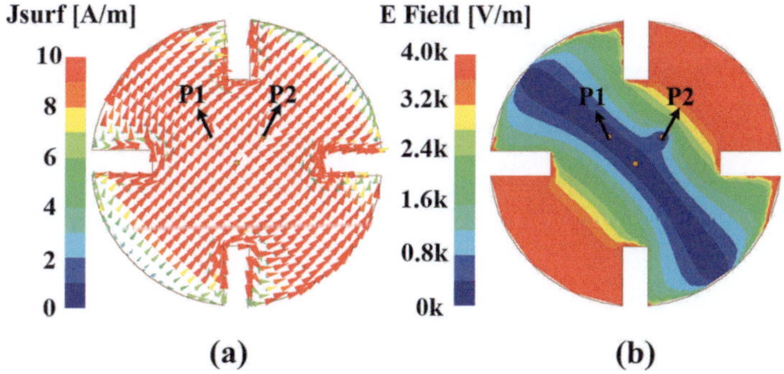

Fig. 3.4 Simulated **a** electric current vector and **b** electric field magnitude distribution at 1.5 GHz on the radiating element when excited independently by the port 1 (the port 2 is terminated with a 50-Ω matched load)

To equip the radiating element with capability of quad-polarization operation, a reconfigurable feeding network is designed on the opposite side of the ground plane. As shown in Fig. 3.2, the substrate of the feeding network is made of F_4BM220 material with thickness of h_2. The specific layout of the reconfigurable feeding network is illustrated in Fig. 3.5a. It mainly consists of two parts: a DC bias circuit with two input ports (i.e., P_{v1} and P_{v2}) and a 3-port radio-frequency (RF) network (with input port P_{in} and output ports P_{o1} and P_{o2}) integrated with four PIN diodes (i.e., D1, D2, D3 and D4) serving as the switches. Considering the insertion loss and cost, the BAR64-02V PIN diodes are chosen and the equivalent circuit model for the OFF and ON states are depicted in Fig. 3.5a. The DC bias circuit is made of two microstrip lines (i.e., the TL_{V1} and TL_{V2}) with width of w_3 and a series of lumped elements, including two 50-pF capacitors and two 1-nH inductors for DC and RF isolation. By controlling the voltages V_1 and V_2 to switch on or off the PIN diodes, the feeding network can operate in four reconfigurable states as shown in Fig. 3.5b.

To clarify the operating principle of the reconfigurable feeding network, the states 1 and 3 are extensively investigated as examples and the other states are not discussed due to similarity. Provided that the PIN diode D1 is switched on and D2 to D4 are off, the reconfigurable feeding network would operate in the state 1 and its equivalent circuit model is shown in Fig. 3.6a. With the RF signal flowing from the input port P_{in} into the output port P_{o1}, the equivalent circuit consists of four transmissions lines (named TL_2, TL_3, TL_5 and TL_6, respectively) with characteristic impedance of Z_{50} (or 50 Ω) and one line with (named TL_4) with characteristic impedance of $Z_{35.36}$ (or 35.36 Ω). Considering that the open-ended parallel branches TL_3 and TL_5 are with the length of one-eighth guided wavelength and the serial branch TL_4 is with the length of one quarter guided wavelength, the input impedances Z_b and Z_{a2} at the nodes b and a can be calculated as

$$Z_b = \frac{Z_{50} \cdot (-jZ_{50})}{Z_{50} - jZ_{50}} = \frac{1-j}{2} \cdot Z_{50} \tag{3.1}$$

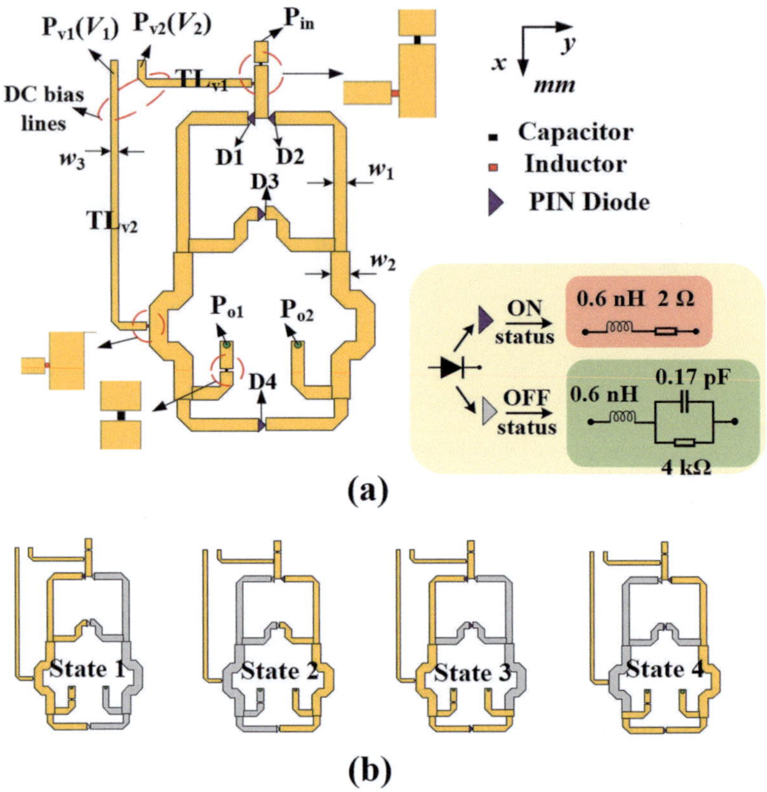

Fig. 3.5 **a** layout of the reconfigurable feeding network and **b** four reconfigurable operation states as controlled by the PIN diodes. ($w_1 = 2.35$ mm, $w_2 = 3.45$ mm, $w_3 = 1.54$ mm, $\lambda_{g1} = 36.5$ mm, $\lambda_{g2} = 36$ mm, $r_2 = 0.6$ mm)

$$Z_{a2} = \frac{Z_{35.36}^2}{Z_b} = \frac{Z_{50}^2}{2Z_b} = \frac{1+j}{2} \cdot Z_{50} \tag{3.2}$$

Then, the input impedance Z_{a1} at the node b can be further derived as

$$Z_{a1} = \frac{Z_{a2} \cdot (-jZ_{50})}{Z_{a2} - jZ_{50}} = Z_{50} \tag{3.3}$$

Obviously, impedance matching is achieved at the input port P_{in}.

Figure 3.6b shows the equivalent circuit model of the reconfigurable feeding network when operating in the state 3. In this situation, the PIN diodes D1, D3 and D4 are switched on and D2 is off, and the RF signal fed from the port P_{in} flow into the ports P_{o1} and P_{o2}. Since the diode D2 is off, the transmission line TL_2' turns to be an open-ended quarter-wavelength line. Accordingly, the node e can then be considered as a virtual ground, which leads to that the lines TL_3 and TL_4' becomes short-ended

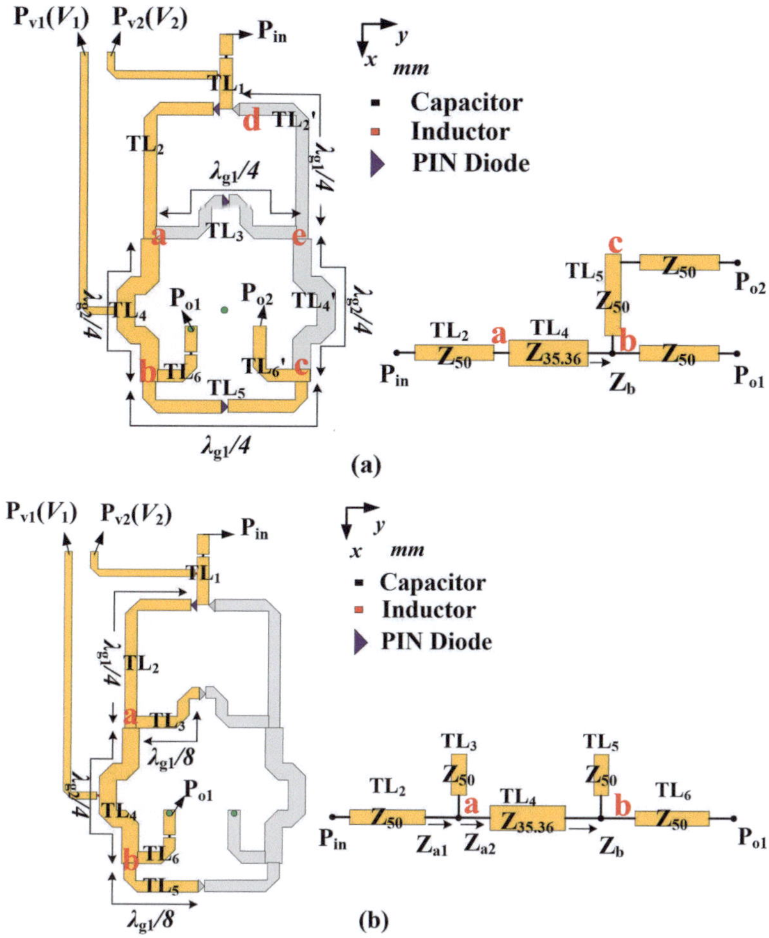

Fig. 3.6 Equivalent circuit models of the reconfigurable feeding network for two different operating states: **a** state 1 and **b** state 3 ($\lambda_{g1} = 36.5$ mm, $\lambda_{g2} = 36$ mm)

quarter-wavelength lines. In this case, virtual open conditions are achieved both at the nodes a and c for the lines TL_3 and TL_4', respectively. Therefore, the feeding network resembles a T-junction 3-dB power divider, as shown in Fig. 3.6b. Besides, the branch TL_4 behaves as a quarter-wavelength impedance transformer for guaranteeing good impedance matching at the node a or input port P_{in}. In addition, since the branch TL_5 has a length of one quarter guided wavelength, the output signals at the ports P_{o1} and P_{o2} have a phase difference of 90°. Due to the circuit symmetry, the feeding network operating in the states 2 and 4 would exhibit performance as similar as that for the states 1 and 3, respectively. Table 3.1 lists the configuration of PIN diodes for all operating states of the reconfigurable feeding network.

Table 3.1 Pin-diodes configuration for all operation states of the reconfigurable feeding network

	D1	D2	D3	D4
State1	On	Off	Off	Off
State2	Off	On	Off	Off
State3	On	Off	On	On
State4	Off	On	On	On

Figure 3.7 shows the simulated scattering parameters of the reconfigurable feeding network when operating in the states 1 and 3. For convenience, the ports P_{in}, P_{o1} and P_{o2} are denoted as 1, 2 and 3, respectively. As shown in Fig. 3.7, the reflection coefficient S_{11} for both states are lower than − 15 dB at the frequencies around 1.5 GHz. For operating state 1, the transmission coefficients S_{21} is above − 0.95 dB within a frequency range from 1.45 to 1.55 GHz, while the S_{31} is below − 15 dB over the band of interest. Seen from Fig. 3.7b, the transmission coefficients S_{21} and S_{31} for state 3 are about − 3.3 dB and − 3.6 dB at the frequency around 1.5 GHz, respectively. Besides, the phase difference between the output signals at the ports P_{01} and P_{02} are around 90° over the operating band.

By integrating the reconfigurable feeding work with the dual-feed miniaturized circular patch radiator, we have simulated the surface current distribution at 1.5 GHz on the polarization reconfigurable radiating element when operating in the states 1 and 3. It can be seen from Fig. 3.8a that, the electric currents only appear on the patch radiator and the right part of the feeding network. Moreover, the magnitudes of the currents on the patch achieve maxima at the moments of 0 and $T/2$, while turning to be minima at the moments of $T/4$ and $3T/4$. The directions of the current vectors on the patch radiator clearly indicates a 45° LP radiation. Figure 3.8b shows the simulated surface current distribution for the operating state 3. Clearly, the electric currents fed from the port P_{in} flows into both the output ports P_{o1} and P_{o2}. Moreover, the current

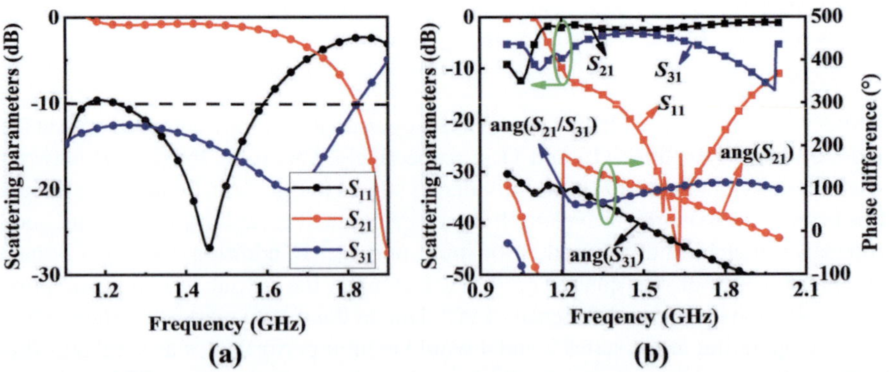

Fig. 3.7 Simulated magnitudes and phases of the scattering parameters for the reconfigurable feeding network when operating in **a** state 1 and **b** state 3

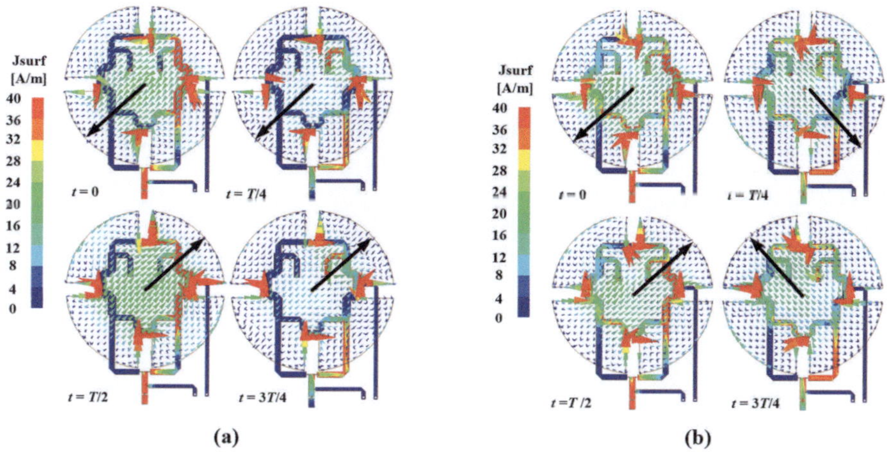

Fig. 3.8 Simulated surface current distribution at 1.5 GHz on the polarization reconfigurable radiating element when operating in the states 1 and 3: **a** 45° LP mode and **b** RHCP mode

vectors on the radiator exhibits a counterclockwise rotation with the variation of time, which indicates a RHCP radiation mode.

3.1.1.2 Design of UAV Conformal Sparse Array Antenna

Provided a large-scale natural disaster occurs, the affected region might face widespread signal outages. The fixed-wing UAV is often used as the emergency communication platform due to its advantages of relatively longer endurance, multiple capabilities and high flexibility. For fixed-wing UAVs such as Wing Loong, the middle part of the drone fuselage can be approximated as a cylinder with radius of 500 mm. It should be mentioned that such approximation causes very minor errors for low frequency antennas, which brings much convenience in UAV conformal array design and modeling.

Faced with various challenges like path loss, multipath fading and unstable connectivity for UAV emergency communications, the UAV communication antenna is desired to feature relatively high gain, polarization diversity and wide range of beam scanning angles. Taking into considerations the aerodynamic characteristics of UAVs as well as the restrictions on the payload weight and cost, a quad-polarization reconfigurable conformal sparse array of high gain and wide coverage range is designed for UAV communication systems. As depicted in Fig. 3.9, the proposed conformal sparse array is attached to the surface of a cylinder with radius of r_c, where nine radiating elements form a 3 × 3 conformal subarray. For sparseness design, three nine-element conformal subarrays are uniformly spaced and arranged on the cylinder bottom. By arranging three subarrays with angular separation of 60° as shown in Fig. 3.9, a relatively wide scanning range of 50° × 180° can be acquired for the entire

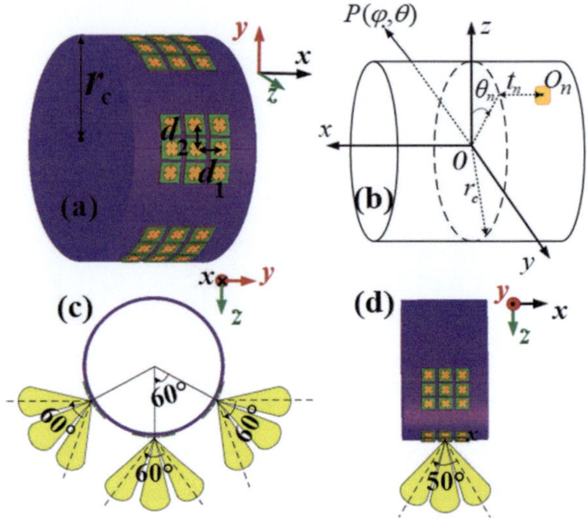

Fig. 3.9 **a** Simulation model of the cylindrically conformal sparse array. **b** Coordinate system for calculating the required phase relations for the array elements. **c** Schematic showing the beam scanning range in the *yz*-plane. **d** Schematic showing the beam scanning range in the *xz*-plane ($d_1 = 0.6\lambda_0$, $d_2 = 0.52\lambda_0$, $r_c = 2.5\lambda_0$)

conformal array through electromagnetic simulation and optimization, which ensures a relatively large coverage area for UAV-based air-to-ground communications.

For each conformal subarray, it is necessary to calculate the required phase relations for all radiating elements. As shown in Fig. 3.9b, a global coordinate system is established with the cylinder center as its origin. The parametric equation for the positions of the radiating elements can be described as

$$\begin{cases} x = t_n \; (-d_1 \le t_n \le d_1, 0 \le \theta_n \le 2\pi) \\ y = r_c \sin \theta_n \\ z = r_c \cos \theta_n \end{cases} \tag{3.4}$$

where r_c is the radius of the cylinder, and t_n and θ_n are the axial and angular positions of the *n*th radiating element, respectively. Hence, the position vector of the radiating element can be denoted as

$$\mathbf{OO_n} = (t_n, r_c \sin \theta_n, r_c \cos \theta_n) \tag{3.5}$$

On the other hand, the position vector of any far field point $P(\theta, \varphi)$ under the spherical coordinate system is described as

$$\mathbf{r_{op}} = (\sin \theta \cos \varphi, \sin \theta \sin \varphi, \cos \theta) \tag{3.6}$$

Then, the difference between the wave path from the source point O_n to the field point P and that from the origin to P is

$$d_n = \mathrm{OO}_\mathbf{n} \cdot \mathbf{r_{op}} \tag{3.7}$$

Hence, the required phase compensation for the nth radiation element can be obtained as

$$\Delta\theta_n = -2\pi d_{\mathrm{n}}/\lambda \quad (3.8)$$

Figure 3.10 shows the simulated realized gains and sidelobe levels of the conformal subarray at 1.5 GHz when beam steering in the *xz*- ($\varphi = 0°$) and *yz*- ($\varphi = 90°$) planes for the − 45°LP and RHCP modes. Clearly, the gain values for both polarization modes fall gradually with increasing the beam steering angle, where the gain variation is less than 1 dBi for beam steering from − 30° to + 30°. Also, the sidelobe levels deteriorate with beam steering, which exceed − 10 dB for both polarization modes at the scanning angles beyond ± 25° and ± 30° in the *xz*-and *yz*-planes, respectively. Figure 3.11a shows the simulated ARs for the RHCP mode with the variation of the beam steering angle, which are less than 1 dB over beam steering from − 30° to + 30° in both planes. Seen from Figs. 3.10 and 3.11a, it is noticed that the radiating performance of the conformal array change more gradually for beam steering in the *yz*-plane than that in the *xz*-plane. Such phenomenon mainly benefits from the wide-angle scanning characteristics of the conformal antennas. Finally, as shown in Fig. 3.11b, the port isolation between adjacent radiating elements are all higher than 20 dB over the frequency band of interest. The simulation analyses for other polarization modes indicate very similar results, which are not shown for brevity.

Based on a comprehensive analysis of the realized gain, sidelobe level and AR performance, it is concluded that the conformal subarray exhibits a satisfactory radiating performance over a beam steering range of 50° × 60° for all polarization modes.

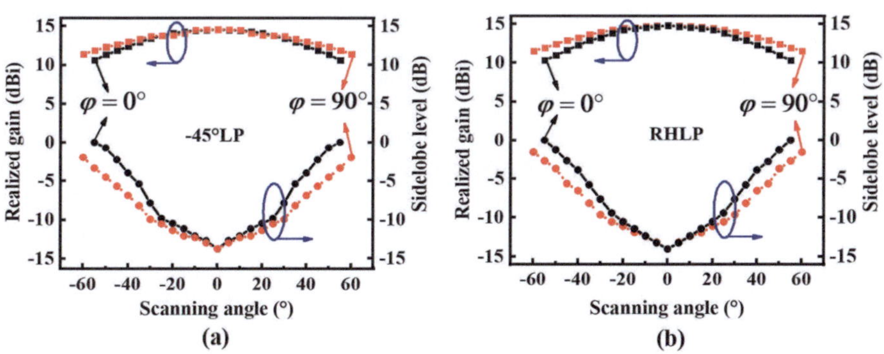

Fig. 3.10 Simulated realized gains and sidelobe levels of the conformal subarray at 1.5 GHz when beam steering in the xz ($\varphi = 0°$) and yz ($\varphi = 90°$) planes for **a** − 45°LP and **b** RHCP modes

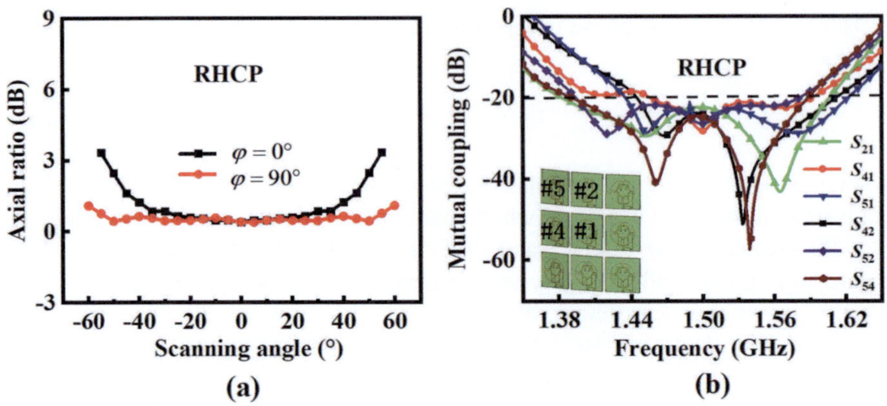

Fig. 3.11 **a** Simulated axial ratios of the conformal subarray at 1.5 GHz when beam steering in the xz ($\varphi = 0°$) and yz ($\varphi = 90°$) planes and **b** mutual coupling between the adjacent radiating elements

3.1.2 Simulation and Measurement

3.1.2.1 Simulation and Measurement of the Radiating Element

Figure 3.12 shows the photograph of the fabricated polarization reconfigurable radiating element, which has a total size of 97 mm × 99 mm. For convenience, the RIGOL DP832 DC power supply was utilized to provide the required biasing voltages for the PIN diodes. As shown in Fig. 3.13 and Table 3.2, configuring different values for the bias voltages V_1 and V_2 can lead to the − 45° LP, + 45°LP, RHCP and LHCP polarization modes of the radiating element, respectively. For all polarization modes, the radiating performance including the reflection coefficients, realized gains, radiation patterns and ARs wherever applicable were measured by using the vector network analyzer and standard horn antenna inside a far-field anechoic chamber at Beijing Institute of Technology.

Figure 3.14 shows the simulated and measured reflection coefficients of the radiating element for different polarization operation modes. Due to the structural symmetry of the feeding network and antenna, the reflection coefficients for two orthogonal LP or CP modes are very similar. As shown in Fig. 3.14a, the measured impedance bandwidth for S_{11} less than − 10 dB is around 55 MHz (i.e., 1.475 to 1.53 GHz) for two LP operation modes, which is 5 MHz wider than the simulated result. On the other hand, it is seen from Fig. 3.14b that the antenna has an impedance bandwidth of 75 MHz (i.e., 1.465–1.54 GHz) for both CP modes, which is also 5 MHz wider than the simulated bandwidth (i.e., 1.47–1.54 GHz). The measured and simulated results are in good agreement, and some minor difference are believed to result from the inaccurate dielectric constant as well as fabrication and assembly errors. Taking into consideration all reconfigurable polarization modes, the antenna has an overlapped impedance bandwidth of 3.66%.

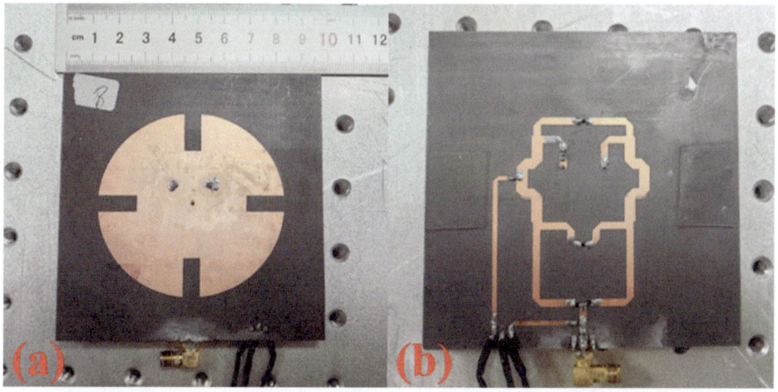

Fig. 3.12 Photograph showing the **a** front and **b** back sides of the polarization reconfigurable radiating element

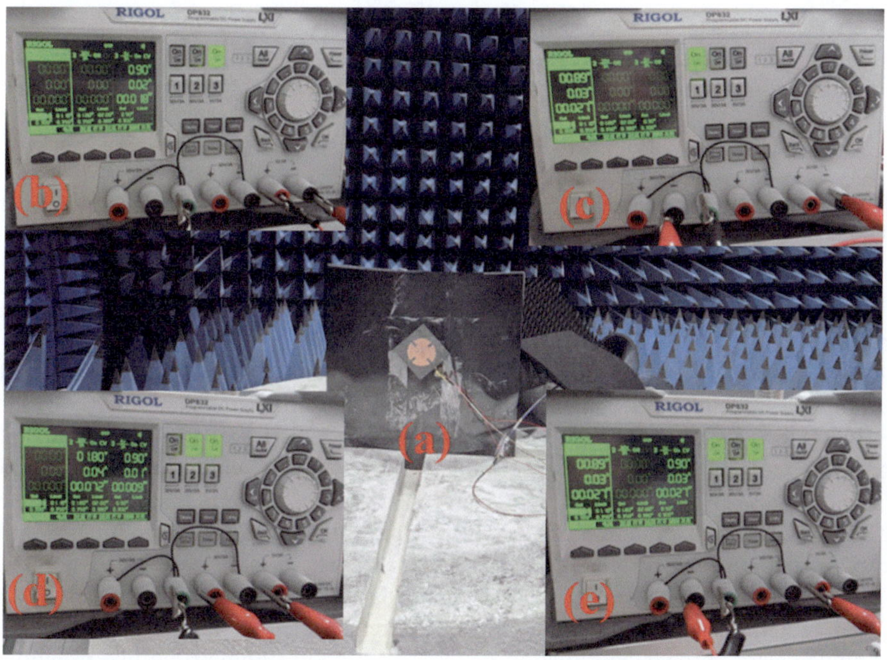

Fig. 3.13 The antenna under test in the measurement chamber. **a** The antenna on the curved plate and dc voltage configurations for **b** − 45°LP, **c** + 45°LP, **d** RHCP and **e** LHCP

Table 3.2 Bias voltage configuration for different polarization modes

Polarization	V_1 (V)	V_2 (V)
− 45°LP	0.9	0
+ 45°LP	− 0.9	0
RHCP	1.8	0.9
LHCP	− 0.9	0.9

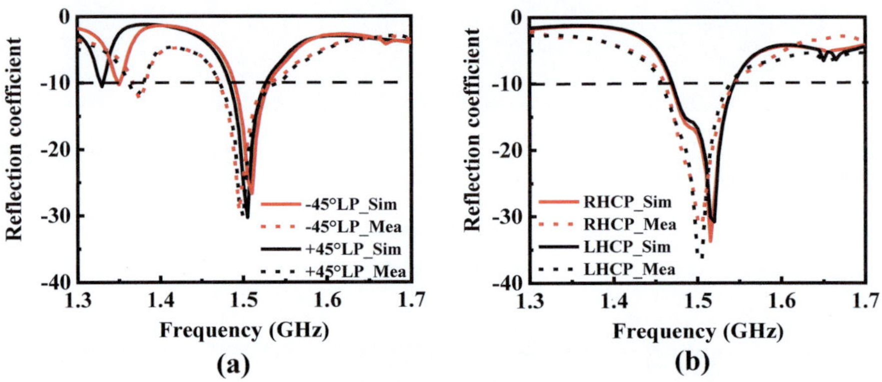

Fig. 3.14 Simulated and measured reflection coefficients of the radiating element for **a** LP and **b** CP modes

Figure 3.15 shows the characterized gain and AR performance for all polarization modes. As shown in Fig. 3.15a, the measured peak gain values are 5.9 dBi for − 45° LP and 6 dBi for + 45° LP modes, while the simulated results are 6.12 dBi and 6.15 dBi, respectively. Seen from Fig. 3.14b, the measured peak gains are 6 dBi for RHCP and 6.1 dBi for LHCP modes, which are 0.2 dB lower than the simulated ones. Moreover, the measured and simulated 3-dB AR bandwidths are 1.485–1.505 GHz (around 1.3%) for both CP modes. It is acceptable that there is some difference between the simulated and measured gain values due to the connection losses and experimental uncertainties.

Figure 3.16 illustrates the simulated and measured radiation patterns at 1.49 GHz for all reconfigurable polarization modes. The results clearly show that the antenna exhibits desired broadside radiation patterns with good symmetry in two principal planes. Moreover, the half-power beam widths (HPBWs) in those two principal planes are both around 70° for all reconfigurable polarization modes. It should be mentioned that the relatively wide beam width of the radiating element helps to reduce beam steering losses for the conformal phased array. As shown in Fig. 3.16, the cross-polarization levels are less than − 15 dB for LP and − 17 dB for CP modes, respectively. The front-to-back ratio is higher than 16 dB for all LP and CP operation modes.

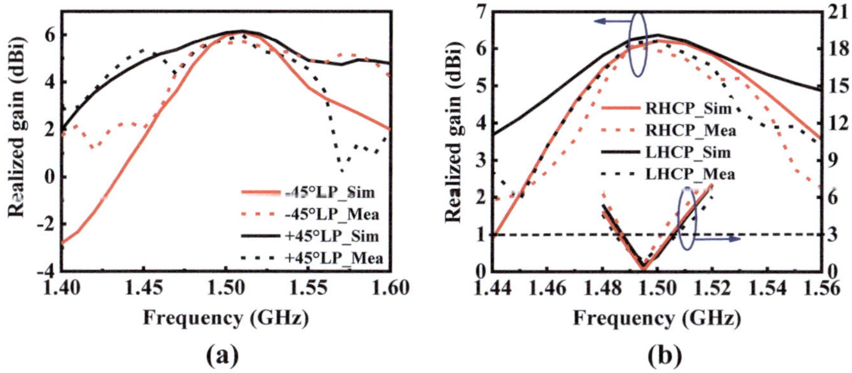

Fig. 3.15 Simulated and measured results for the realized gain and AR of the radiating element for **a** LP and **b** CP modes

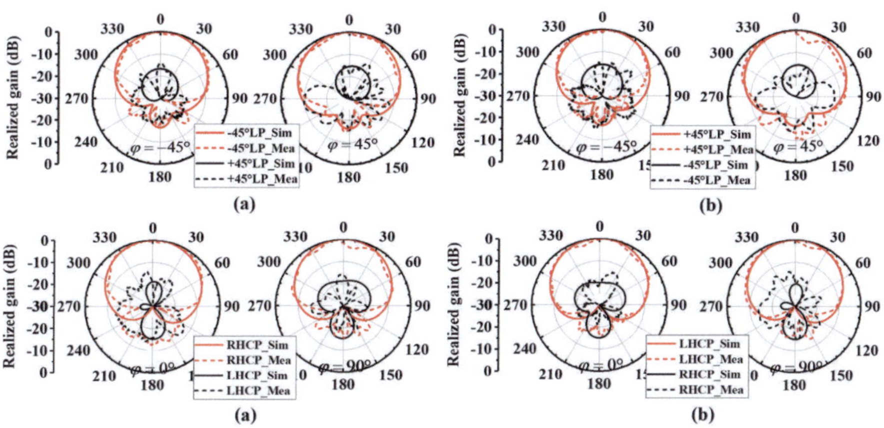

Fig. 3.16 Simulated and measured radiation patterns of the radiating element at 1.49 GHz for **a** − 45° LP, **b** + 45° LP, **c** RHCP and **d** LHCP operation modes

Table 3.3 provides a performance comparison for the proposed radiating element with other reported state-of-the-art low-profile polarization reconfigurable antennas. It is noted that the proposed antenna occupies a smallest area (i.e., $0.485\lambda_0 \times 0.495\lambda_0$) as compared with other reported antennas except for the Huygens dipole antenna°, thus beneficial to improving the radiating characteristics of the conformal array. Apart from the antenna reported in Ref. [1], the proposed antenna integrates the fewest PIN diodes and DC biasing lines, enabling four reconfigurable polarization modes while significantly reducing the loss and network complicity. The proposed antenna exhibits the highest radiation efficiency (i.e., 84%) for all polarization modes. In addition, the proposed antenna makes a good compromise between the bandwidth and gain performance. Although the antenna reported in Ref. [2] exhibits a much wider bandwidth and higher gain, its thickness and area are far larger than that of the proposed antenna,

Table 3.3 Simulated IMBWs And ARBWs of four antenna elements

	Ant. 1 (%)	Ant. 2 (%)	Ant. 3 (%)	Ant. 4 (%)
IMBW	22.8	31.8	34	32.8
ARBW	5.45	19.9	19.3	12.8

thus not very suitable for conformal array applications. Overall, the proposed radiating element exhibits superior comprehensive performance as compared with other state-of-the-art antennas.

3.1.2.2 Simulation and Measurement of the Conformal Array

To characterize the radiating performance of the polarization reconfigurable conformal array, three 9-way power dividers with different fixed phase shifts are designed and fabricated for facilitating the measurement process. As shown in Fig. 3.17, utilizing three phase-shift power dividers helps to measure the radiating characteristics of the conformal array under three beam steering angles, i.e., (φ, θ) = (0°, 0°), (0°, 25°) and (90°, 30°), respectively. Three phase-shift power dividers were measured and they all have exhibited good performance. To avoid overlength of the paper, only the simulation and measured results of one power divider are presented. Figure 3.18 illustrates the simulated and measured scattering parameters of the 9-way power divider with the phase shifts corresponding to the beam steering angle of ($\varphi = 0°$, $\theta = 25°$). The reflection coefficient S_{11} is less than − 20 dB in the frequency range of 1.45–1.55 GHz. Moreover, the transmission coefficients S_{21} to S_{101} are all around − 10 dB over the whole operating frequency band. Figure 3.18c and d shows the desired phase relationship for nine radiating elements, thus guaranteeing that the main beam of the conformal array steers to the angle of ($\varphi = 0°$, $\theta =$ 25°). By connecting the other phase-shift power dividers to the radiating elements, the conformal array can generate directional radiations towards different angles.

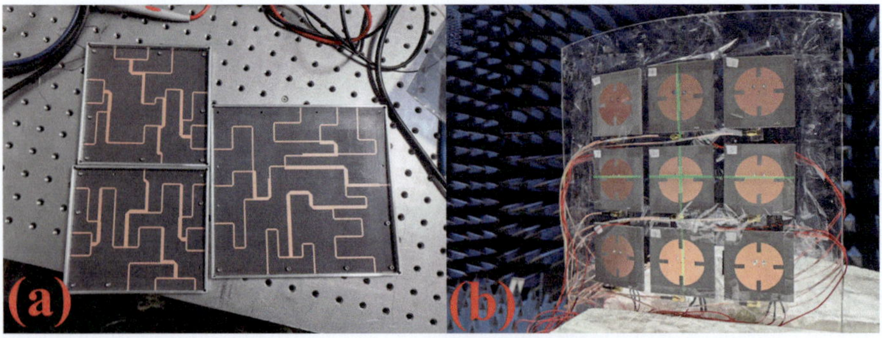

Fig. 3.17 Photographs of fabricated prototype of the proposed phase shifters and patch array antenna. **a** the 1:9 phase shifters. **b** the 3 × 3 conformal patch array antenna

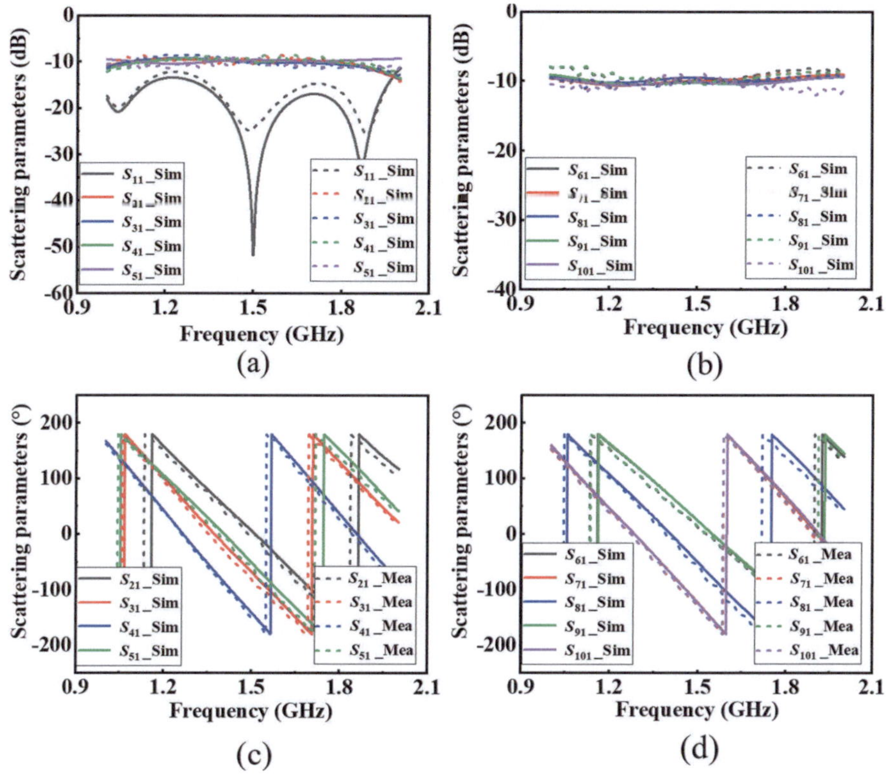

Fig. 3.18 Simulated and measured scattering parameters of one 9-way power divider with the phase shifts corresponding to the beam steering angle of ($\varphi = 0°$, $\theta = 25°$): **a** the magnitudes of S_{11} to S_{51}; **b** the magnitudes of S_{61} to S_{101}; **c** the phases of S_{21} to S_{51}; **d** the phases of S_{61} to S_{101}

One 3 × 3 conformal subarray was characterized in a microwave anechoic chamber, as shown in Fig. 3.17b. Considering the similarity and the length limit of the paper, only the simulated and measured results for the − 45° LP and RHCP modes are provided.

Figure 3.19 shows the simulated and measured reflection coefficients of the conformal subarray integrated with the 9-way power divider for those three beam directions. Since it would consume too much memory and time for performing a full-wave analysis of the combined antenna system, the reflection coefficients are simulated by utilizing the software ANSYS High Frequency Structure Simulator (HFSS) and Advanced Design System (ADS). The proposed simulation method starts with acquiring all scattering parameters of the conformal array and the 9-way power divider independently from HFSS, then performs a circuit or network simulation of the combined antenna system modelled in ADS. As shown in Fig. 3.19a, the simulated overlapped impedance bandwidth for $S_{11} < -10$ dB is around 111 MHz (i.e., 1.474–1.585 GHz) for the − 45° LP mode for three beam steering angles, which is slightly narrower than the measured result (i.e., 1.443–1.567 GHz). For the RHCP operation mode, it is seen from Fig. 3.19b that the simulated overlapped bandwidth is around 116 MHz (i.e., 1.462–1.578 GHz), which is also narrower than the measured one (i.e., 1.442–1.572 GHz). Overall, the simulated and measured curves are in reasonably well agreement, which have demonstrated enough operating bandwidth of the conformal subarray. Some minor difference might be caused by the material losses as well as the fabrication and measurement errors.

Figure 3.20 shows the simulated and measured realized gains and ARs of the conformal subarray for those three beam directions. As shown in Fig. 3.20a, the measured peak gains for the − 45° LP mode are approximately 13.64 dBi, 14.18 dBi and 13.77 dBi for the beam steering angle of (0°, 25°), (0°, 0°) and (90°, 30°), respectively, which are 0.3 dBi lower than the simulated results. Seen from Fig. 3.20b, the measured peak gains for the RHLP mode are around 13.63 dBi, 14.35 dBi and

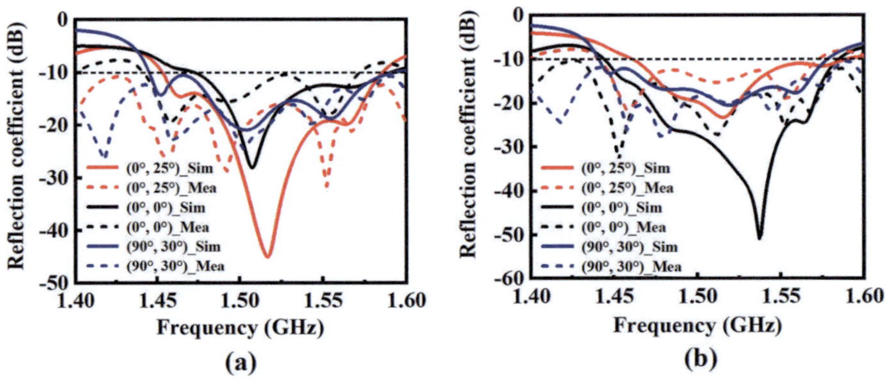

Fig. 3.19 Simulated and measured reflection coefficients of the conformal subarray for three beam directions for **a** − 45° LP and **b** RHCP modes

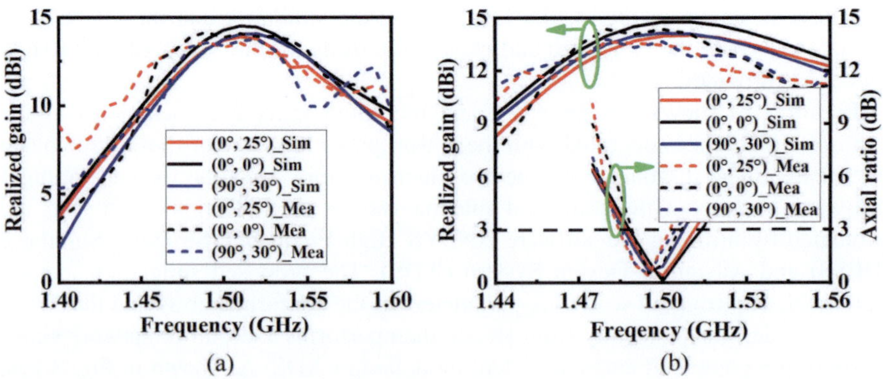

Fig. 3.20 Simulated and measured realized gains and ARs of the conformal subarray for three beam directions for **a** − 45° LP and **b** RHCP modes

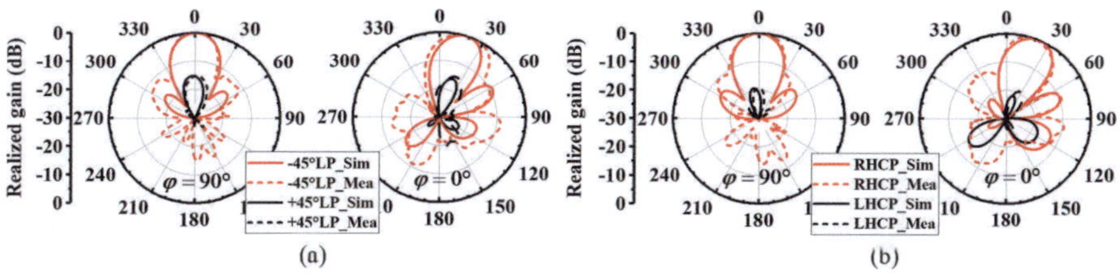

Fig. 3.21 Simulated and measured radiation patterns of the conformal subarray at 1.49 GHz for **a** – 45° LP and **b** RHCP operation modes when the desired beam direction at ($\varphi = 0°$, $\theta = 25°$)

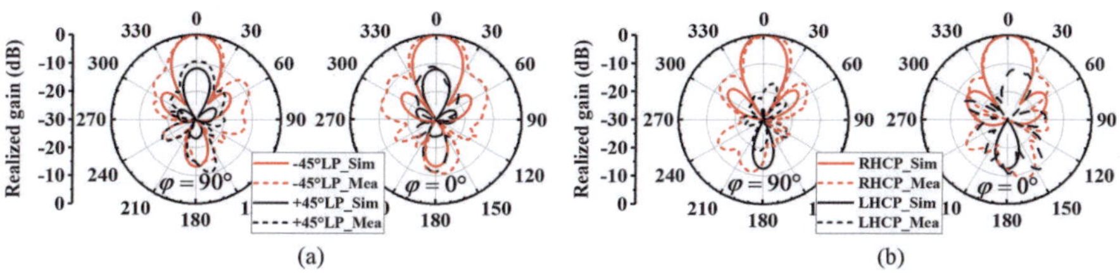

Fig. 3.22 Simulated and measured radiation patterns of the conformal subarray at 1.49 GHz for **a** – 45° LP and **b** RHCP operation modes when the desired beam direction at ($\varphi = 0°$, $\theta = 0°$)

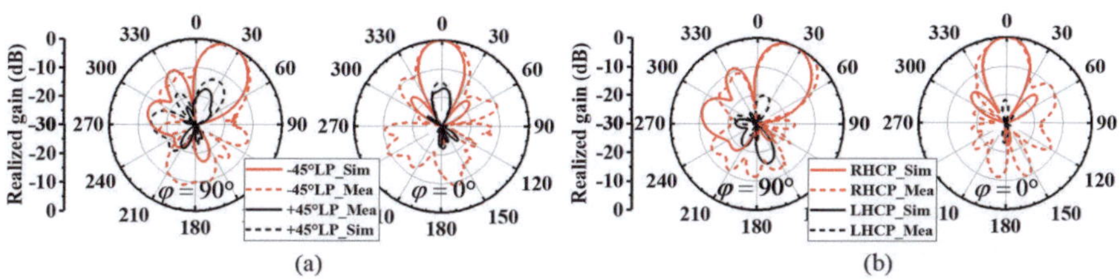

Fig. 3.23 Simulated and measured radiation patterns of the conformal subarray at 1.49 GHz for **a** – 45° LP and **b** RHCP operation modes when the desired beam direction at ($\varphi = 90°$, $\theta = 30°$)

13.79 dBi for three beam directions, while the simulated results are 14 dBi, 14.77 dBi and 14.14 dBi, respectively. In addition, the measured 3-dB AR bandwidths are around 1.3% for all beam steering angles, where the frequency range moves downward by around 5 MHz as compared with the simulated result. Such minor difference might be caused by the fabrication and measurement errors.

Figures 3.21, 3.22 and 3.23 show the simulated and measured radiation patterns of the polarization reconfigurable conformal array at 1.49 GHz for those three beam directions, respectively. It is clearly seen that for both the – 45° LP and RHCP operation modes, the conformal array radiates directional beams toward the desired angles of (0°, 25°), (0°, 0°) and (90°, 30°), respectively. It is observed that the measured HPBW of the conformal array in the two principal planes are around 30° for two polarization modes. Besides, the cross-polarization levels are lower than – 15 dB and – 18 dB for the – 45° LP and RHCP modes, respectively. Due to the disturbance of the DC biasing lines and the assembly errors for array elements in experiment, relatively higher sidelobe and back lobe levels are observed for the measured radiation patterns as compared with the simulated results. Such difference can be minimized by employing more advanced integration and assembly technologies. Overall, the measured and simulated radiation patterns are in good agreement for different polarization modes and beam steering angles.

3.2 A Helmet-Based Hemispherical-Conformal Wearable Array Antenna with Broad Bandwidth, Wide-Angle Coverage and Reconfigurable Dual Circular Polarizations for WBAN Applications

This section presents a helmet-based hemispherical conformal wearable array antenna with broad bandwidth, wide-angle coverage and reconfigurable dual CPs for WBAN applications. By utilizing a single-feed polarization reconfigurable crossed dipole incorporated with parasitic patches above a ground, a broadband conformal radiating element with unidirectional radiations is designed. The element integrated with PIN-diodes can flexibly operate in two orthogonal CP modes by switching the on–off states of the diodes. To broaden the operating bandwidth, four sector shaped patches working as parasitic radiators are loaded on four quadrants of the crossed dipole. Besides, a 12-element helmet-based hemispherical-conformal wearable array antenna with two reconfigurable CP modes is designed, which is arranged in special configuration for improving the gain, coverage range and orientation flexibility for WBAN systems. The experimental results show that the proposed hemispherical-conformal wearable array exhibits an overlapped 3-dB ARBW of 29.6% and − 10-dB IMBW of 69.8% at around 3 GHz for both reconfigurable CP modes. Besides, a peak realized gain more than 8 dBi and a beam coverage range of $110° \times 360°$ are obtained. The superior radiating performance of the proposed antenna has proved its huge potential for wearable WBAN applications.

3.2.1 *Antenna Design and Analysis*

3.2.1.1 Design of Broadband Unidirectional Dual-CPs Reconfigurable Conformal Radiating Antenna Element

Since wearable antennas are usually integrated onto clothing or helmets, effectively reducing the electromagnetic scattering or absorption by the human body is a crucial consideration. Hence, a unidirectional radiating element with high front-to-back ratio is desired for wearable antenna designs. To attain CP unidirectional radiation, one pair of crossed dipoles with different arm lengths are designed above a finite ground plane. For the initial design named Ant. 1 as shown in Fig. 3.24a, two adjacent arms of the crossed dipole are connected to the inner conductor of the coaxial feeding port, and the other arms are connected to the ground plane via shorting pins. Considering the distribution of the resonant electric currents on the half-wave crossed dipole, the radiation field of the antenna can be decomposed into two orthogonal LP modes. To obtain CP radiation, differentiating the dipole length is an effective way to impose perturbation as well as generate the desired relationship of equal magnitude and 90° phase difference between those two orthogonal LP radiation modes. By optimizing the unidirectional CP radiation pattern for wearable applications, the height of the crossed dipole from the ground plane is finalized to be around one quarter effective wavelength at the central frequency (i.e., 3 GHz). However, due to the presence of a ground plane close to the crossed dipole, whether the impedance bandwidth (IMBW) or axial ratio bandwidth (ARBW) of Ant. 1 are very narrow, as shown in Fig. 3.25. To broaden the operating bandwidth, four sector-shaped patches working as parasitic elements are placed in the four quadrants of the crossed dipole, then Ant. 1 turns to be Ant. 2. By co-optimizing the dimensions of the crossed dipole and parasitic patches, both the IMBW and ARBW of Ant. 2 are significantly larger than those of Ant. 1 (see Fig. 3.3 and Table 3.1). To enable reconfigurable dual-CPs radiation capability, Ant. 2 is further upgraded to Ant. 3 by integrating four PIN diodes at optimized positions along four arms of the crossed dipole, as illustrated in Fig. 3.24c. By controlling the bias voltage to generate a $\pm$ 90° phase difference between two orthogonal LP modes, a reconfigurable LHCP or RHCP mode can thus be achieved. After size optimization, Ant. 3 exhibits dual-CPs reconfigurable radiation characteristics with the bandwidth performance as similar as that of Ant. 2, as shown in Fig. 3.25 and Table 3.3. Lastly, to analyze performance variation from planar to curved configuration, a hemispherical-conformal version (i.e., Ant. 4) with curvature of $K = 8.33$ is designed while keeping other geometric parameters unchanged as compared with Ant. 3. It is clearly seen that Ant. 4 exhibits an IMBW as similar as that of Ant. 3, but its ARBW is a little smaller due to structural bending.

Based on Ant. 4, a broadband dual-CPs reconfigurable conformal radiating element that suitable for the helmet is designed as shown in Fig. 3.26. As shown

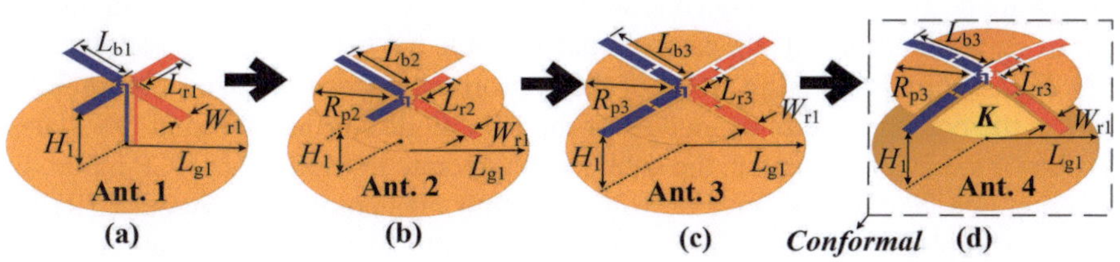

Fig. 3.24 Operating principle and evolution of the broadband unidirectional dual-CPs reconfigurable radiating element: **a** unidirectional crossed-dipole antenna; **b** broadband unidirectional antenna; **c** broadband unidirectional dual-CPs reconfigurable antenna; **d** conformal broadband unidirectional dual-CPs reconfigurable antenna ($L_{b1} = 19$ mm, $L_{r1} = 15$ mm, $L_{g1} = 33$ mm, $H_1 = 20$ mm, $W_{r1} = 4$ mm, $L_{b2} = 28$ mm, $L_{r2} = 14$ mm, $R_{p2} = 25$ mm, $L_{b3} = 28$ mm, $L_{r3} = 9$ mm, $R_{p3} = 23$ mm, $K = 8.33$)

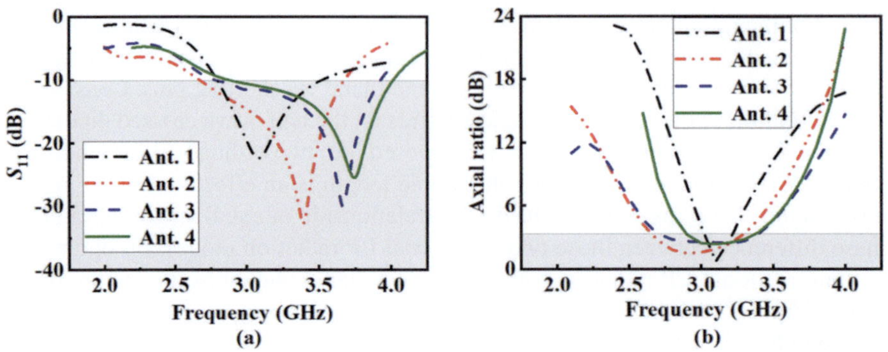

Fig. 3.25 **a** Simulated reflection coefficients and **b** ARs with the frequency for Ant. 1, Ant. 2, Ant. 3 and Ant. 4

in Fig. 3.26a, the antenna element is mainly composed of three parts: substrate, radiating part and ground plane. Considering that the radiating element is conformal to the hemispherical helmet, an arch-shaped cylinder with curvature K and height H_s is designed for the substrate. The material of the substrate is Acrylonitrile Butadiene Styrene (ABS) with dielectric constant of 3.2, which is lightweight and easy to be molded. The conformal radiator located on top layer of the substrate consists of the crossed dipole and parasitic patches. As compared with the feeding part of Ant. 4, one additional shorting pin is utilized for connecting the adjacent harms to the ground plane, which effectively reduces the parasitic inductance introduced by one single shorting pin. To flexibly switch the operating mode of the crossed dipole, four BAR64-02V PIN diodes (i.e., D1, D2, D3 and D4) with operational frequency range from 10 MHz to 6 GHz, are placed at Ld1 away from the open ends of the arms, respectively. The dc bias points for the PIN diodes are located near the open arm ends by connecting a 37-nH inductor for RF isolation. Besides, one shorting pin with radius of Rs is utilized for connecting one parasitic patch and the ground plane for dc bias grounding. As shown in Fig. 3.26a, the positive terminal of D1 and the negative terminal of D2 are directly grounded through the shorting pins, while D3 and D4 are dc grounded through one shorting pin and two inductors that bridging one pair of adjacent arms and one parasitic patch. Table 3.4 lists the configuration of the PIN diodes for all operating states of the conformal antenna element. Figure 3.27a depicts the equivalent circuit models of the PIN diode for the OFF and ON states. As shown in Fig. 3.27b and c, the antenna can operate in Model 1 (LHCP mode) by applying one identical positive dc voltage at four bias points, and Model 2 (RHCP mode) by applying one negative dc voltage. It should be emphasized that regardless of dc grounding, only one bias line is required due to the identical dc voltage applied at four bias points.

To illustrate the CP radiation capability, the surface electric current distribution on the conformal radiating element is simulated. As shown in Fig. 3.28, when the antenna operates in the LHCP mode, the sum vector for the surface currents on whether the crossed dipole or parasitic patches rotates clockwise in one period. Notably, since

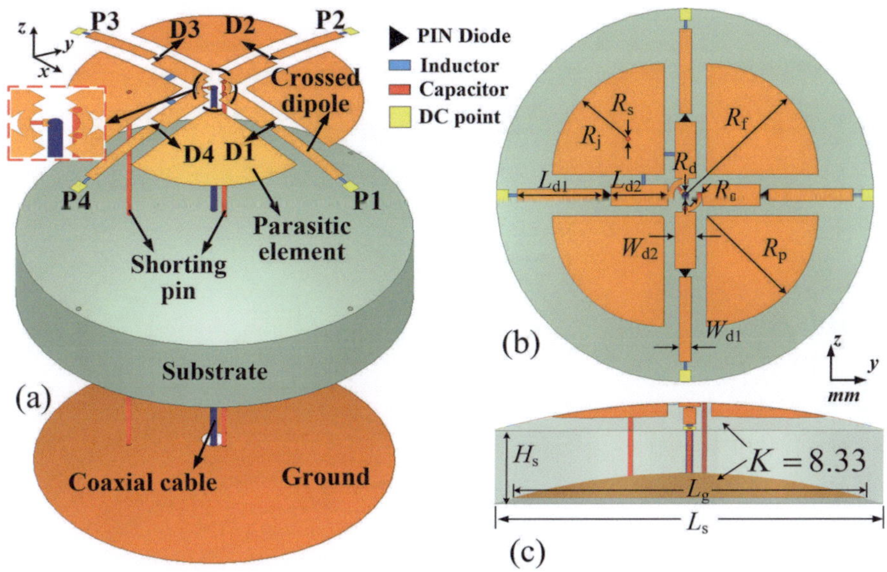

Fig. 3.26 Geometrical configuration of the broadband dual-CPs reconfigurable conformal radiating element: **a** 3D-view; **b** top view; **c** side view (Rs = 0.4 mm, Rj = 14 mm, Rf = 23.2 mm, RP = 19.5 mm, Rd = 0.6 mm, Ld1 = 15 mm, Ld2 = 10 mm, Wd1 = 2.1 mm, Wd2 = 3.7 mm, Hs = 12 mm, Ls = 66 mm, Lg = 60 mm)

Table 3.4 Pin-diodes configuration for all operation states of the conformal antenna element

	D1	D2	D3	D4
LHCP	On	Off	On	Off
RHCP	Off	On	Off	On

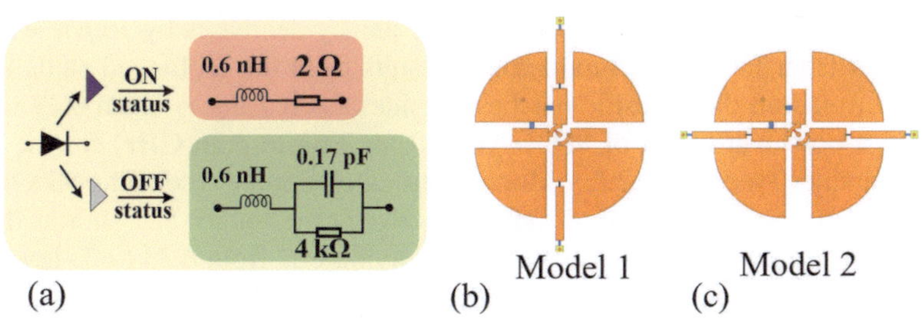

Fig. 3.27 **a** Equivalent circuit models of the BAR64-02V PIN diode operating in the ON and OFF states; **b** reconfigurable LHCP and **c** RHCP modes for the conformal radiating element

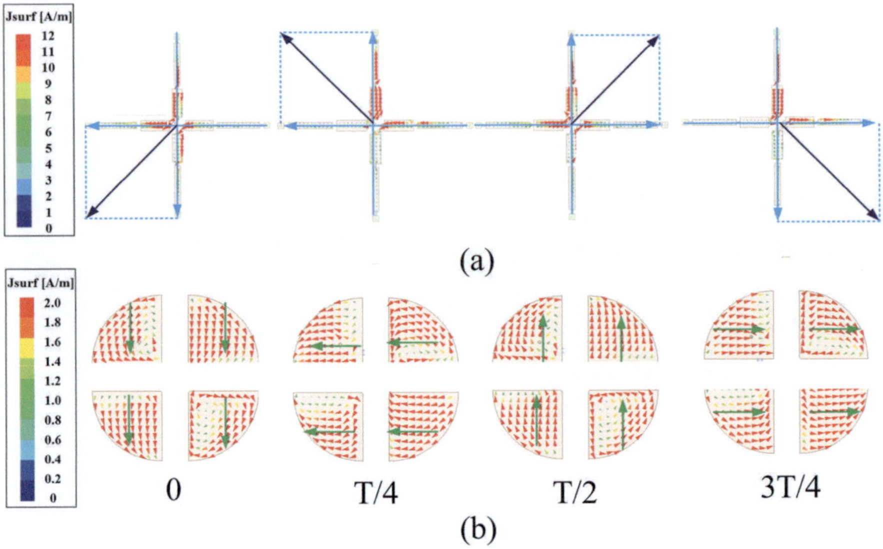

Fig. 3.28 Simulated surface current distributions on the crossed dipole and parasitic patches of the conformal radiating element when operating in the LHCP mode at the frequency of 3 GHz

the shorting pin connecting the ground to the parasitic patch is located at one quarter wavelength away from the edge of the parasitic patch, the position of the shorting pin is equivalent to a virtually short point, thus causing very little disturbation on the current mode on the conformal antenna. Considering the similarity, the surface current distribution for the RHCP mode is not described and discussed.

Figures 3.29 and 3.30 show the impacts of some key parameters (i.e., R_p, $r_L = L_{d2}/L_{d1}$, K and H_s) on radiation performance of the conformal radiating element. As shown in Fig. 3.29a, the highest operating frequency decreases by 13.6% when R_p is increased from 17.5 to 23.5 mm, which results from the prolonged paths for the electric currents on the parasitic patches. Besides, increasing r_L from 0.47 to 0.79 results in a downward shift of 8.33% (i.e., from 3.61 to 3.36 GHz) for the upper cutoff frequency (see Fig. 3.29b). More specifically, double resonant peaks tend to disappear with too large length difference between the longer and shorter dipoles (i.e., smaller r_L). In addition, increasing the curvature K from 8.33 to 20 lowers the operating band by 4.5% (see Fig. 3.30a), which is mainly caused by the changed capacitances and inductances in the equivalent resonant circuit of the conformal radiating element while being bent to different extent. It is observed from Figs. 3.29 and 3.30a that as compared with the reflection coefficient, the AR is more sensitive to the variations of geometric parameters, which strongly depends on the amplitude and phase relationships between two orthogonal linear polarization modes on the antenna. Figure 3.30b illustrates the *xz*-plane radiation patterns at 3 GHz, showing that increasing the curvature reduces the peak gain by 0.97 dB (i.e., from 5.65 to 4.68 dBi) and broadens the half-power beam width by 10° (i.e., from 94° to 104°),

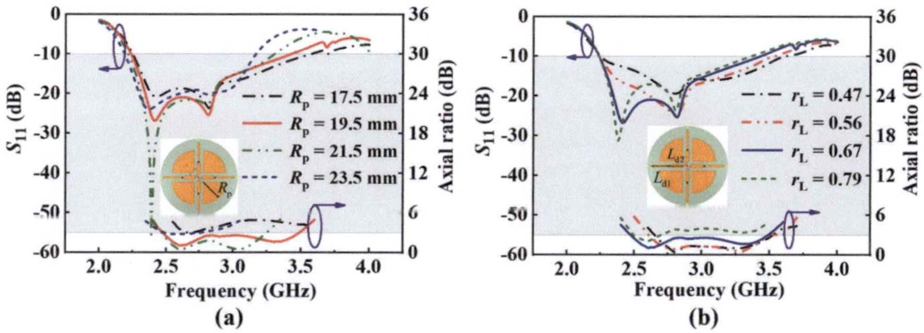

Fig. 3.29 Simulated reflection coefficients and ARs of the conformal radiating element with the frequency for **a** different R_p and **b** different r_L

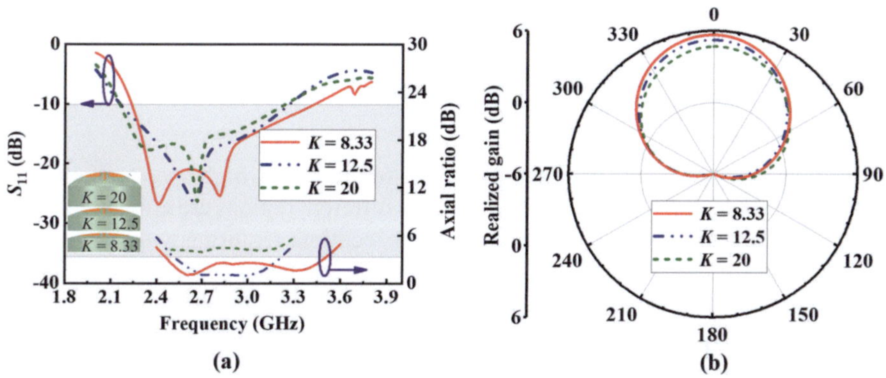

Fig. 3.30 Simulated reflection coefficients, ARs and *xz*-plane radiation patterns (at 3 GHz) of the conformal radiating element for different *K*

which can readily be understood from the decrease in the projected effective radiation aperture. Lastly, Fig. 3.31 illustrates the influence of the height Hs on the reflection coefficient, AR and realized gain. It is found that both the S_{11} and AR curves shift downward with increasing H_s, which indicates the presence of the ground plane can cause the variation of frequency response of the antenna. With a height optimized to be around one quarter effective wavelength at the central frequency (i.e., $H_s = 12$ mm), the conformal radiation element exhibits a peak gain of 5.67 dBi, showing a 1.44 dB enhancement as compared with the case of $H_s = 16$ mm. Taking as accounts the actual helmet radius and antenna performance, the curvature K and height H_s are finalized to be 8.33 and 12 mm, and R_p and r_L are optimized to be 19.5 mm and 0.67, respectively. The simulated − 10-dB IMBW and 3-dB ARBW of the conformal radiating element are 2.24–3.42 GHz and 2.5–3.45 GHz, respectively.

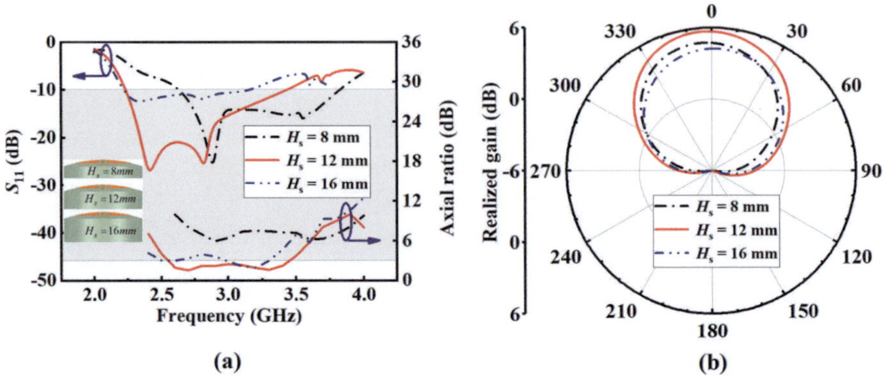

Fig. 3.31 Simulated reflection coefficients, ARs and *xz*-plane radiation patterns (at 3 GHz) of the conformal radiating element for different H_s

3.2.1.2 Design of Helmet-Based Hemispherical-Conformal Dual-CPs Reconfigurable Wearable Array

In the emergency rescue scenarios typically with forest fires, high-speed reliable and stable communications between the rescuers in different regions are critically important for guaranteeing various cooperative task execution, requiring wide bandwidth, reconfigurable polarization as well as sufficient gain and beam coverage range for the wearable antennas. Here, a hemispherical-conformal dual-CPs reconfigurable wearable array consisting of 12 elements is designs on a helmet with radius of 120 mm. As shown in Fig. 3.32, the conformal array antenna is comprised of three elements on the top and nine elements surrounding the side bottom. To improve the realized gain, the top three elements work simultaneously to form a fixed beam with an optimized beam width of 40° × 40°. Similarly, two adjacent elements at the side bottom work together to increase the gain, creating a beam width of 90° × 40°. Hence, an overall beam coverage range of $(\theta, \varphi) = (110° \times 360°)$ can be achieved by electrically switching between the top three elements and the side bottom arbitrary adjacent two elements. In this way, the effective beam coverage zone of the proposed helmet-based hemispherical-conformal wearable array antenna can cover more than the upper half space, as illustrated in Fig. 3.32.

For beamforming synthesis, it is necessary to calculate the required phase relations for all relevant radiating elements. As shown in Fig. 3.33a, a global coordinate system is established with the spherical center as its origin. The parametric equation for the positions of conformal radiating elements can be described as

$$\begin{cases} x = r_h \sin\theta_n \cos\varphi_n \\ y = r_h \sin\theta_n \sin\varphi_n \\ z = r_h \cos\theta_n \end{cases} \tag{3.9}$$

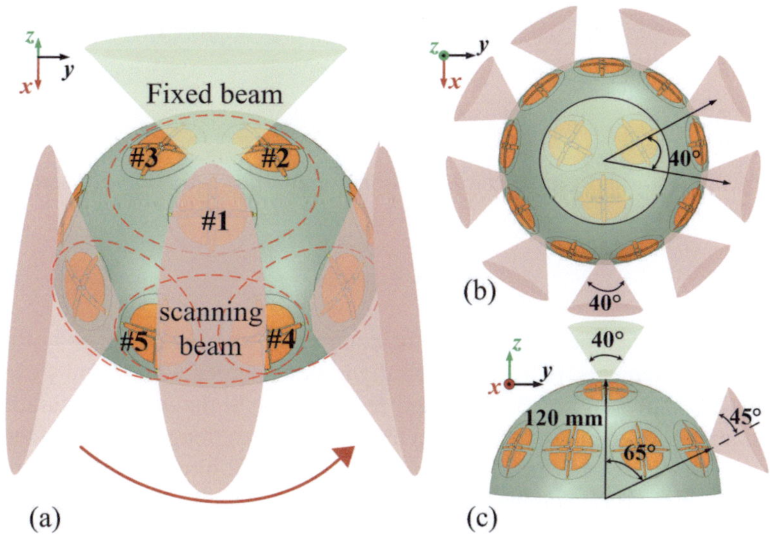

Fig. 3.32 Structural geometry of the hemispherical-conformal dual-CPs reconfigurable wearable array antenna

Fig. 3.33 **a** Coordinate system for calculating the required phase relations for the array elements; **b** and **c** top views of two configurations for the top three radiating elements

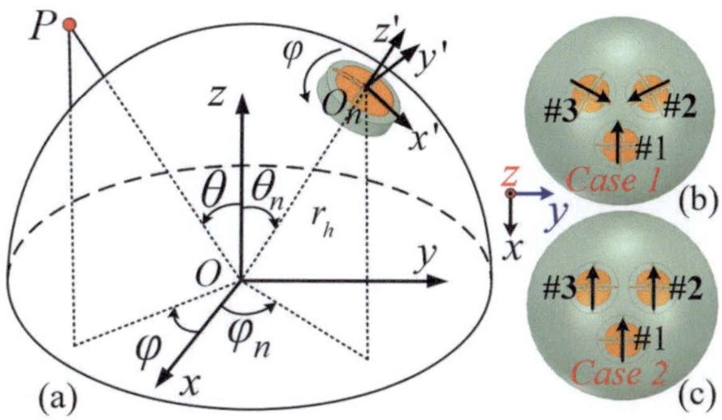

where r_h is the radius of the hemisphere, and φ_n and θ_n are the azimuth and pitch angles of the nth radiating element, respectively. Hence, the position vector of the nth conformal radiating element can be denoted as

$$\mathbf{OO_n} = (r_h \sin\theta_n \cos\varphi_n, r_h \sin\theta_n \sin\varphi_n, r_h \cos\theta_n) \tag{3.10}$$

On the other hand, the normalized position vector of any far-field point $P(\theta, \varphi)$ under the spherical coordinate system is described as

$$\mathbf{r_{op}} = (\sin\theta \cos\varphi, \sin\theta \sin\varphi, \cos\theta) \tag{3.11}$$

Then, the difference between the distance from the source point O_n to the field point P and that from the origin O to P is

$$d_n = \mathrm{OO_n} \cdot \mathbf{r_{op}} \tag{3.12}$$

Hence, the required phase compensation for the nth radiation element can be obtained as

$$\Delta\theta_n = -2\pi d_\mathrm{n}/\lambda \tag{3.13}$$

To achieve optimal performance of the conformal subarray comprised of the top three radiating elements, two feasible configurations named *Case 1* and *Case 2* are considered as shown in Fig. 3.33b and c, respectively. In the first configuration, all elements are arranged in a rotationally symmetric manner along the z axis of the main coordinate system, where the angle between the normal direction of each element and + z-axis is θ_n. Based on the first configuration, a second configuration as shown in Fig. 3.33c can be obtained by rotating the antenna elements #2 and #3 along the z'-axis of their respective local coordinate systems as shown in Fig. 3.33a. For either LHCP or RHCP operation mode, all those radiating elements are exactly the same in terms of their structures and PIN-diode states. To clearly illustrate the difference between those two configurations, a dark arrow along the projected longer active dipole on the *xy*-plane is marked in Fig. 3.33b and c. Obviously, the adjacent arrows have a separation angle of 120° for the first configuration, while all arrows are along the -*x*-axis direction for the second configuration. Figure 3.34 shows the simulated realized gains, ARs and *xz*-plane radiation patterns at 2.8 GHz for two configurations of the top three-element subarray when operating in the LHCP mode. It is noted that although the realized gains and radiation patterns are almost identical for those two configurations, the ARBW for *Case 1* is much wider than that for *Case 2*, with the AR value close to zero over the whole band.

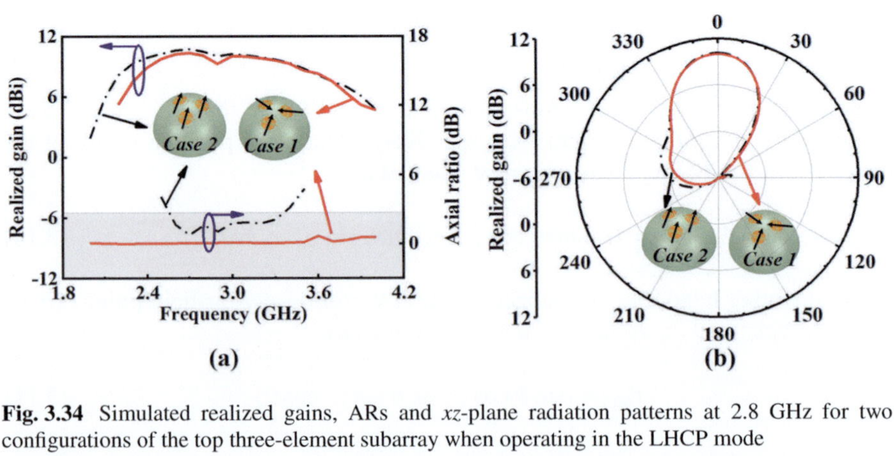

Fig. 3.34 Simulated realized gains, ARs and *xz*-plane radiation patterns at 2.8 GHz for two configurations of the top three-element subarray when operating in the LHCP mode

The reasons for the significant difference in AR performance for two configurations are explained as follows. As shown in Fig. 3.35a, the dark arrow along the projected longer active dipole of the radiating element is along the -x-axis. The radiating electric field vectors are $\mathbf{E}_L$ and $\mathbf{E}_R$ for the LHCP and RHCP modes, respectively, which are both assumed to be along the y-axis direction initially. If the radiating element is rotated clockwise along the z-axis by a, the direction of the field vector will be rotated by the same angle, which is equivalent to introducing a leading or lagging phase a to $\mathbf{E}_L$ or $\mathbf{E}_R$, respectively. Similarly, if there is a feed phase b for the radiating element, the directions of the field vectors will be rotated by the same angle b but in opposite directions for those two CP modes. The consequence is that the phase change or rotation angle is the same for $\mathbf{E}_L$ while opposite for $\mathbf{E}_R$ when a and b get the same value. According to the definition of the AR, the AR value for a CP antenna should satisfy

$$1 \le AR_{\text{ele}} = \frac{|\mathbf{E}_L| + |\mathbf{E}_R|}{|\mathbf{E}_L| - |\mathbf{E}_R|} \le \sqrt{2} \tag{3.14}$$

$$ARBW_{\text{ele}} \in [A, B] \tag{3.15}$$

where AR_{ele} and $ARBW_{\text{ele}}$ refer to the AR and 3-dB ARBW of the radiating element, respectively. Obviously, A and B represent the lowest and highest frequency points for CP radiation.

The electric field vectors $\mathbf{E}_{\text{L-case1}}$ and $\mathbf{E}_{\text{R-case1}}$ for the top three-element subarray configured in *Case 1* are shown in Fig. 3.35b, where the subarray is assumed to operate in its LHCP mode. Obviously, $\mathbf{E}_{\text{L-case1}}$ and $\mathbf{E}_{\text{R-case1}}$ are the sums of the co-polarized (i.e., $\mathbf{E}_{L1}$, $\mathbf{E}_{L2}$ and $\mathbf{E}_{L3}$) and cross-polarized electric field vectors (i.e., $\mathbf{E}_{R1}$, $\mathbf{E}_{R2}$ and $\mathbf{E}_{R3}$) for three radiation elements, respectively. Thus, the AR and 3-dB ARBW of the top three-element subarray can be derived as

$$|\mathbf{E}_{\text{L-case1}}| = |\mathbf{E}_{L1} + \mathbf{E}_{L2} + \mathbf{E}_{L3}| = 3|\mathbf{E}_L| \tag{3.16}$$

$$|\mathbf{E}_{\text{R-case1}}| = |\mathbf{E}_{R1} + \mathbf{E}_{R2} + \mathbf{E}_{R3}| = 0 \tag{3.17}$$

$$AR_{\text{case1}} = \frac{|\mathbf{E}_{\text{L-case1}}| + |\mathbf{E}_{\text{R-case1}}|}{|\mathbf{E}_{\text{L-case1}}| - |\mathbf{E}_{\text{R-case1}}|} \equiv 1 \tag{3.18}$$

$$ARBW_{\text{case1}} = \infty \tag{3.19}$$

where AR_{case1} and $ARBW_{\text{case1}}$ refer to the AR and 3-dB ARBW of the three-element subarray, respectively. As clearly shown in Fig. 3.35b, the field vectors $\mathbf{E}_{L1}$, $\mathbf{E}_{L2}$ and $\mathbf{E}_{L3}$ are exactly the same in both amplitude and phase, while the vectors E_{R1}, E_{R2} and E_{R3} have the same amplitude but specific phase relations, leading to a zero value for $|E_{\text{R-case1}}|$. Consequently, AR_{case1} is calculated to be 1 (or 0 dB) and $ARBW_{\text{case1}}$ is infinitely large. On the other hand, as shown in Fig. 3.35c, when the top three

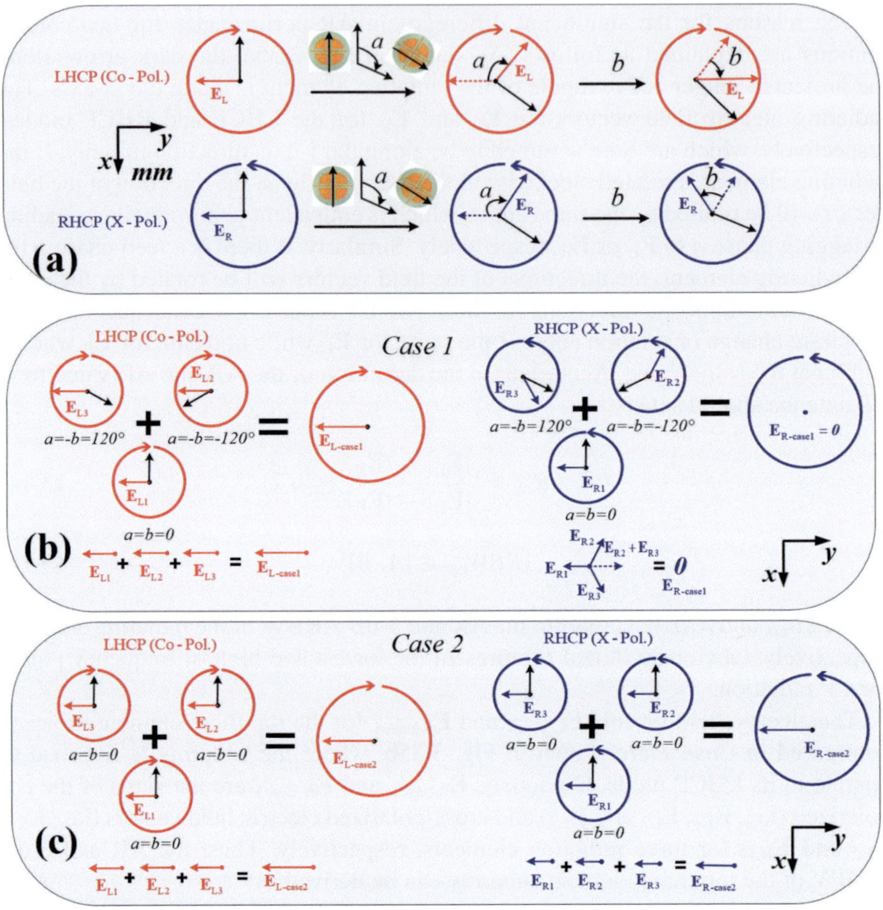

Fig. 3.35 Schematic diagram illustrating the phase and angular relationships between the vector electric fields for **a** the conformal radiating element and **b** and **c** the conformal top three-element subarray configured in *Case 1* and *Case 2*, respectively

radiating elements are configured in *Case 2* and the subarray is still assumed to operate in its LHCP mode. In this case, the AR_{case2} and $ARBW_{\text{case2}}$ can be derived as

$$|\mathbf{E}_{\text{L-case2}}| = |\mathbf{E}_{\text{L1}} + \mathbf{E}_{\text{L2}} + \mathbf{E}_{\text{L3}}| = 3|\mathbf{E}_{\text{L}}| \tag{3.20}$$

$$|\mathbf{E}_{\text{R-case2}}| = |\mathbf{E}_{\text{R1}} + \mathbf{E}_{\text{R1}} + \mathbf{E}_{\text{R1}}| = 3|\mathbf{E}_{\text{R}}| \tag{3.21}$$

$$AR_{\text{case2}} = \frac{|\mathbf{E}_{\text{L-case2}}| + |\mathbf{E}_{\text{R-case2}}|}{|\mathbf{E}_{\text{L-case2}}| - |\mathbf{E}_{\text{R-case2}}|} = AR_{ele} \tag{3.22}$$

$$ARBW_{\text{case2}} \in [A, B] \tag{3.23}$$

It is evident that the ARBW of the subarray operating in *Case 2* is identical to that of the radiating element. It should be emphasized that for whatever array configuration of the three elements, the directions or phases of the co-polarized electric field vectors must be in coincidence for guaranteeing in-phase superposition and thus maximum gain. Based on the analysis, the conformal subarray arranged in *Case 1* is found to exhibit the best AR performance among different configurations of top three radiating elements.

Figure 3.36 illustrates the impacts of θ_n on the AR, realized gain and radiation pattern in the *xz*-plane at 2.8 GHz for the top three-element subarray. It can be seen that the AR value remains essentially the same, while the peak realized gain decreases by 0.37 dB when θ_n increases from 19° to 28°. As shown in Fig. 3.36b, the main-lobe gain decreases while the sidelobe level rises with the increase of θ_n. This phenomenon can be explained as that, the increased separation angle between the normal directions of conformal radiating elements leads to reduced contribution in radiations toward the desired + *z*-axis direction. However, although the subarray for $\theta_n = 19°$ exhibits the highest gain, those three radiating elements are extremely close to each other, even leading to slightly overlapping at their bottom sides. Taking into considerations both the realized gain and assembly complexity, the angle θ_n is finally selected to be 22°.

Apart from the top three conformal radiating elements, there are nine elements uniformly distributed surrounding the side bottom of the hemispherical helmet. of which the normal directions are all at an angle of 65° with the *z*-axis. By sequentially exciting a certain number of adjacent elements, a beam scanning range of 360° in azimuth can be achieved. To determine the number of active elements, performance simulation analysis is carried out for two-, three- and four-element subarrays. As shown in Fig. 3.37, according to the phase relationship as described in Section II-B, the required phase compensations for conformal array elements at specific positions are calculated using Eqs. (2)-(5) for constructive interference and thus increasing the gain toward the desired beam direction. It can be observed that with appropriate phase

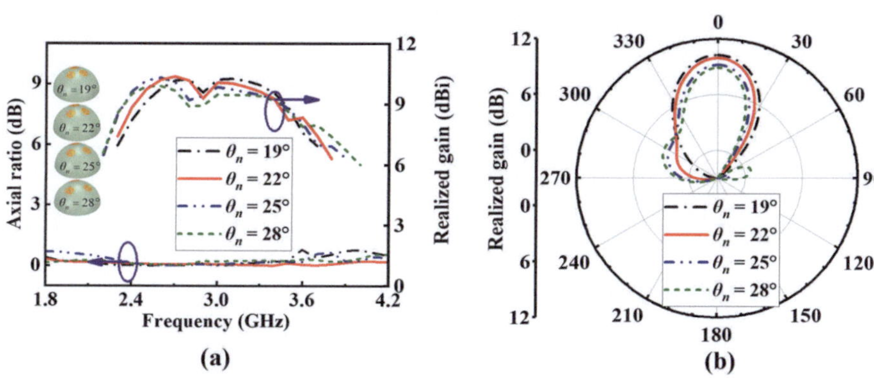

Fig. 3.36 **a** Simulated ARs, realized gains and **b** radiation patterns in the xz-plane at 2.8 GHz for the top three-element subarray when θ_n increases from 19° to 28°

compensation, the peak realized gain of the four-element subarray is 0.92 dB higher than that of the two-element subarray, which results from the increase in the effective radiation aperture. In comparison, the AR performance remains largely unaltered in that such beam synthesis has very little influence on the polarization for the array configuration as shown in Fig. 3.37a. It should be noted that, the half-power beam widths (HPBWs) in the $y''z''$-plane (the new coordinate system x″y″z″ is established by rotating the coordinate system xyz clockwise around the y-axis by 65°) are 24° and 30° for the three-element and four-element subarrays, respectively, which do not meet the requirement of 360° azimuthal coverage by utilizing nine elements surrounding the helmet. In comparison, the HPBWs in two principal planes for the two-element subarray is around 90° × 40° (see Fig. 3.16), thus resulting in an overall beam-steering coverage range of $(\theta, \varphi) = (90° \times 360°)$ via beam switching among nine elements. In addition, the influence of adjacent inactive elements on the radiation performance of two-element subarray is observed to be very little, as shown in Fig. 3.38. Hence, by combining the radiation patterns of the top three-element subarray with one fixed beam and the side bottom nine-element (two elements active) subarray with steerable beam, which allows a wide beam coverage range as shown in Fig. 3.32.

After determining the arrangement of conformal radiating elements on the helmet, it is necessary to examine the influences of operating frequency variation on the HPBWs of the top three-element subarray and the side bottom two-element subarray. As shown in Fig. 3.39a, the HPBWs of the top three-element conformal subarray are wider than 40° over the frequency range of 2.2–3.4 GHz (i.e., the overlapped bandwidth of IMBW and ARBW) in both the xz- and yz-planes. Seen from Fig. 3.39b, the HPBWs for the two-element subarray are wider than 93° in the $x''z''$-plane and 39° in the $y''z''$-plane over the frequencies from 2.5 to 3.4 GHz (i.e., the overlapped bandwidth of IMBW and ARBW), respectively. Consequently, the proposed conformal array combining two types of subarrays can achieve more than half-space coverage range over the whole operating band.

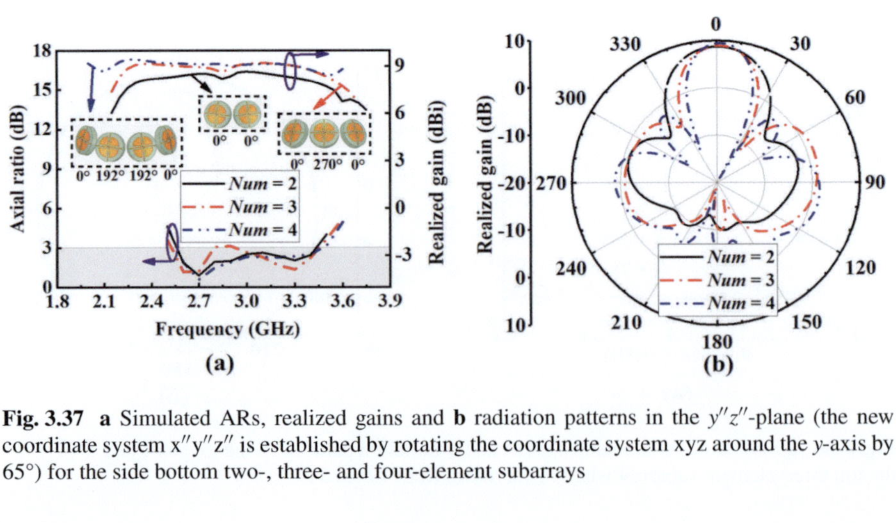

Fig. 3.37 **a** Simulated ARs, realized gains and **b** radiation patterns in the $y''z''$-plane (the new coordinate system x″y″z″ is established by rotating the coordinate system xyz around the y-axis by 65°) for the side bottom two-, three- and four-element subarrays

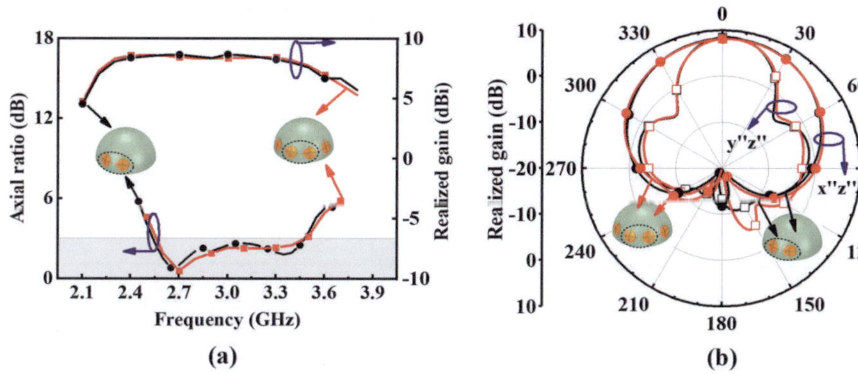

Fig. 3.38 **a** Simulated ARs, realized gains and **b** radiation patterns at 2.8 GHz of the side bottom two-element subarray for the cases of without and with two other adjacent inactive elements

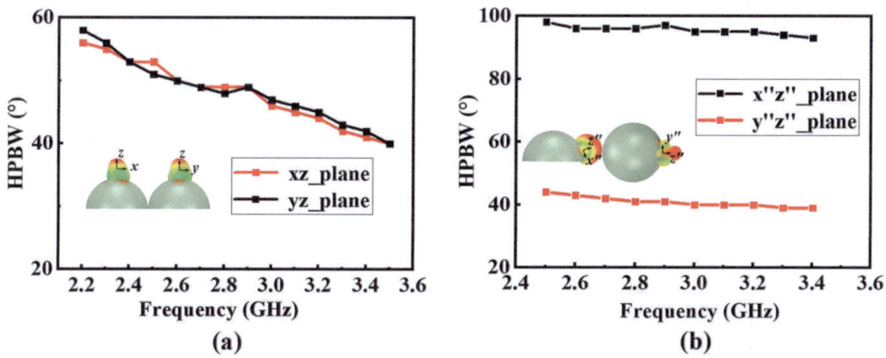

Fig. 3.39 Simulated HPBWs in two principal planes versus the operating frequency for the **a** top three-element subarray and **b** side bottom two-element subarray when operating in the LHCP mode

For conformal array antennas, it is necessary to examine the influence of mutual coupling between the conformal radiating elements in the array [4–6]. According to Ref. [7], the straight-line distance between two points on a spherical surface can be expressed as

$$d_{\text{line}} = \sqrt{2r_{\text{h}}^2[1 - \sin\theta_1 \sin\theta_2 \cos(\varphi_1 - \varphi_2) - \cos\theta_1 \cos\theta_2]} \tag{3.24}$$

As shown in Fig. 3.32, the spacing between element #1($\theta_1 = 22°$, $\varphi_1 = 0°$) and #2($\theta_2 = 22°$, $\varphi_2 = 120°$) in the top three-element subarray is $0.78\lambda_0$ (where λ_0 is the free-space wavelength at 3 GHz). In the side bottom nine-element subarray, the optimized spacing between element #4($\theta_4 = 65°$, $\varphi_4 = -20°$) and #5($\theta_5 = 65°$, $\varphi_5 = 20°$) is $0.74\lambda_0$. Additionally, the distance between element #1 and #5 is $0.91\lambda_0$. The spacing between the phase centers of the top three-element subarray and the side

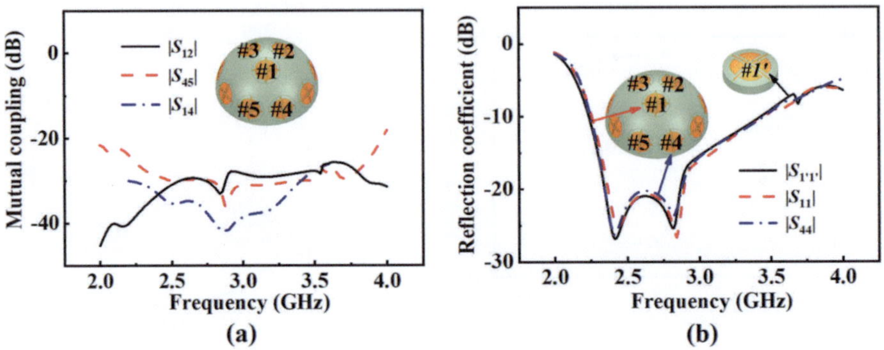

Fig. 3.40 **a** Simulated mutual coupling and **b** reflection coefficients for the conformal radiating elements when operating in the LHCP mode

bottom two-element subarray is $1.28\lambda_0$. As shown in Fig. 3.40a, the simulated port isolation between adjacent radiating elements exceeds 25 dB over the frequency band of interest for the two subarrays operating in the LHCP mode. Furthermore, it can be observed that the isolation between the inter-array elements also remains very high. As shown in Fig. 3.40b, small mutual coupling leads to a negligible difference in S_{11} values for the conformal radiating element working alone and inside the subarrays. Figure 3.41 illustrates the influences of mutual coupling on the radiation patterns of one conformal radiating element of those two conformal subarrays, which clearly shows the presence of other adjacent elements cause very little degradation on its own radiation characteristics of the conformal element. Finally, Fig. 3.42 shows the simulated 3D radiation patterns of two conformal subarrays at different frequencies (i.e., 2.6, 3 and 3.4 GHz) for the LHCP operation mode. It is seen that while gradually rising with the operating frequency, the sidelobe levels are all relatively low thus not influencing the practical applications of the proposed antenna.

3.2.2 *Simulation and Measurement*

3.2.2.1 Simulation and Measurement of the Conformal Radiating Element

Figure 3.43 shows the photograph of the fabricated broadband unidirectional dual-CPs reconfigurable conformal radiating element, which has a total size of $\pi \times 33^2 \times 12\ mm^3$. As illustrated in Fig. 3.43d, the element to be tested was mounted onto a hemispherical structure made of ABS, which has a radius of 120 mm and a thickness of 12 mm. For convenience, a RIGOL DP832 dc power supply was utilized to provide the required biasing voltages for the PIN diodes. With the biasing voltage set to 0.9 and − 0.9 V, the conformal radiating element can operate in the LHCP and RHCP modes, respectively. For both polarization modes, the radiating performance

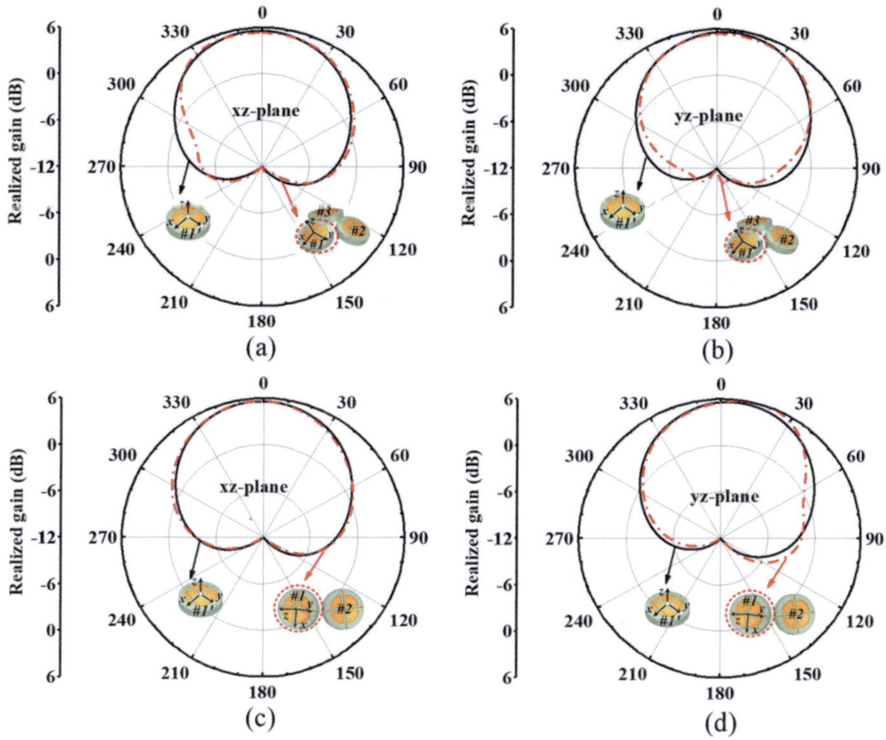

Fig. 3.41 The influences of mutual coupling on the radiation patterns at 3 GHz of one conformal radiating element of the: **a** top three-element subarray in the *xz*-plane; **b** top three-element subarray in the *yz*-plane; **c** side bottom two-element subarray in the *xz*-plane; **d** side bottom two-element subarray in the *yz*-plane. All radiation patterns are analyzed for the LHCP mode

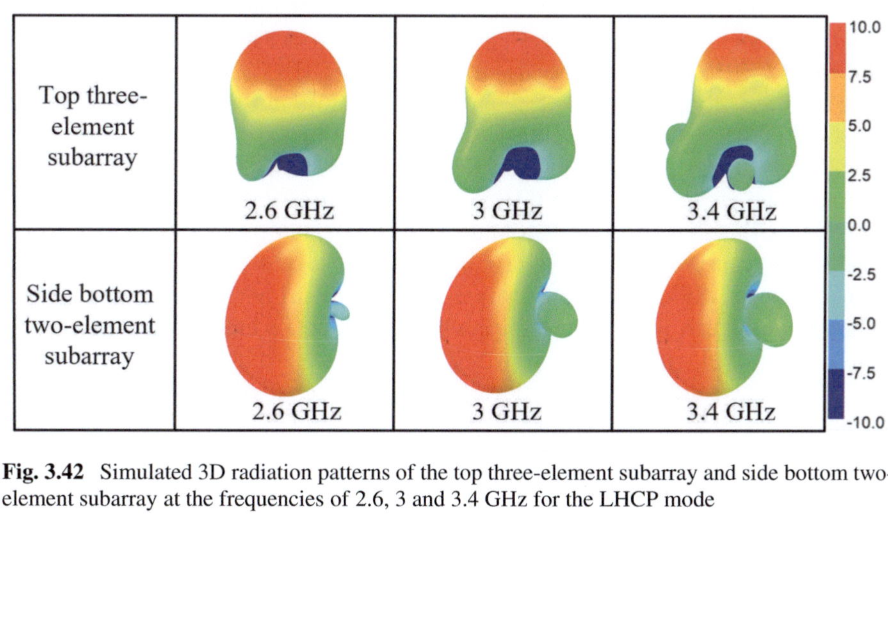

Fig. 3.42 Simulated 3D radiation patterns of the top three-element subarray and side bottom two-element subarray at the frequencies of 2.6, 3 and 3.4 GHz for the LHCP mode

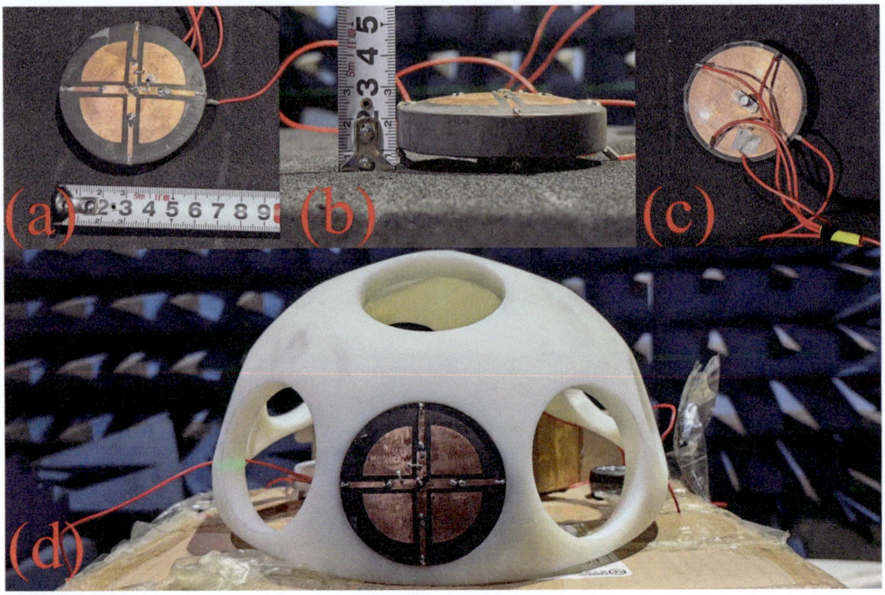

Fig. 3.43 Photograph showing the **a** front, **b** side and **c** back sides of the conformal radiating element and **d** the antenna under test in an anechoic chamber

including the reflection coefficients, ARs, realized gains and radiation patterns, were measured by using a vector network analyzer and a standard horn antenna inside a far-field anechoic chamber at Beijing Institute of Technology.

The simulated and measured reflection coefficients, ARs and realized gains of the conformal radiating element for both polarization modes are plotted in Fig. 3.44. Due to the structural symmetry of the conformal radiating element, the reflection coefficients for two CP modes are very similar. As shown in Fig. 3.44a, the measured overlapped IMBW for S_{11} less than − 10 dB is 40.3% (i.e., 2.31–3.48 GHz) for the LHCP and RHCP operation modes, which is very close to the simulated results of 40.6% (i.e., 2.24–3.42 GHz). Figure 3.44b shows that the measured overlapped 3-dB ARBW is 33% (i.e., 2.52–3.51 GHz) for both modes, which agrees well with the simulated one of 32% (2.5–3.45 GHz). The minor differences between the simulation and experimental results are attributed to the fabrication and measurement errors. Besides, the measured realized gains range from 4.7 to 5.58 dBi over the operating frequency band, while the simulated ones are between 5.1 and 6 dBi for both polarization modes. It can be seen that there is a little bit reduction in the measured realized gains as compared with simulation, which mainly results from the losses associated with PIN diode soldering and nonideal substrate material used for fabricating the antenna.

Figure 3.45 shows the simulated and measured normalized radiation patterns of the conformal radiating element at 2.6, 3, and 3.4 GHz for two reconfigurable CP operation modes. The results show that the HPBWs in two principal planes tend to be

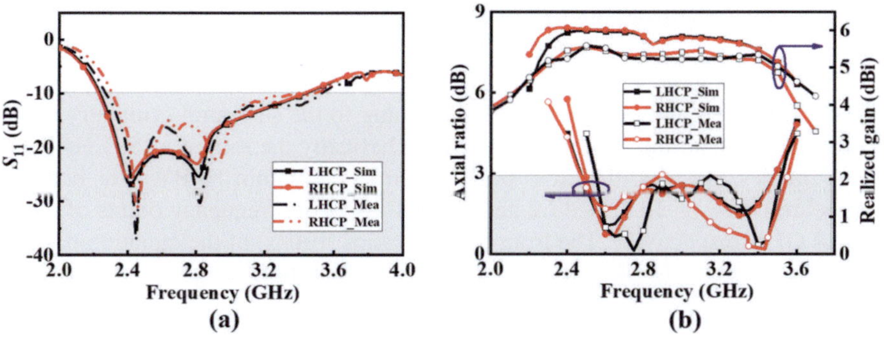

Fig. 3.44 Simulated and measured results for the **a** reflection coefficients, **b** ARs and realized gains of the conformal radiating element operating in the LHCP and RHCP modes

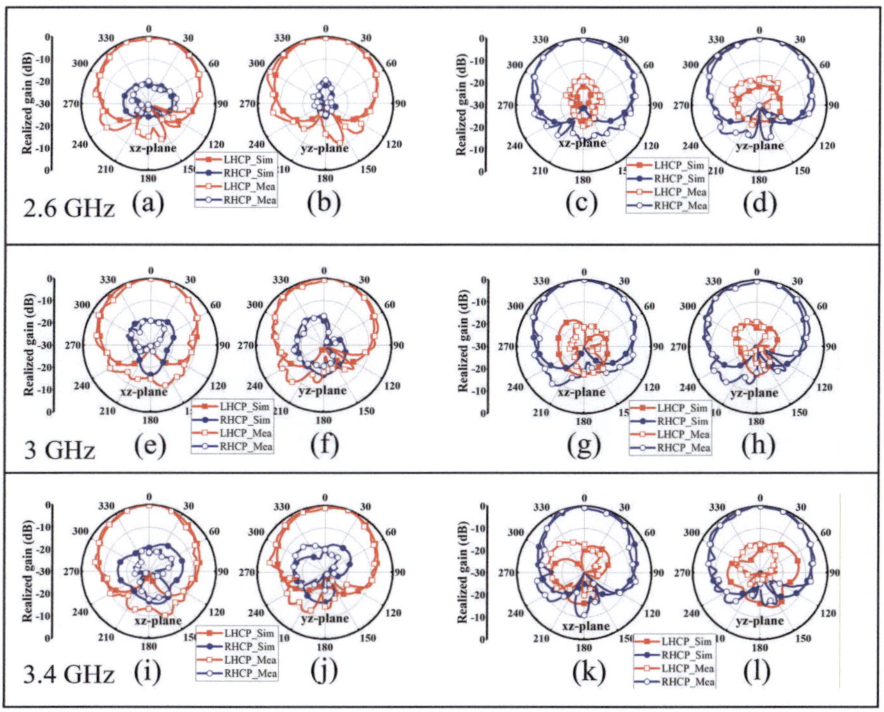

Fig. 3.45 Simulated and measured radiation patterns of the conformal radiating element at different frequencies and different principal planes for different polarization modes: **a** *xz*- and **b** *yz*-planes at 2.6 GHz for LHCP operation; **c** *xz*- and **d** *yz*-planes at 2.6 GHz for RHCP operation; **e** *xz*- and **f** *yz*-planes at 3 GHz for LHCP operation; **g** *xz*- and **h** *yz*-planes at 3 GHz for RHCP operation; **i** *xz*- and **j** *yz*-planes at 3.4 GHz for LHCP operation; **k** *xz*- and **l** *yz*-planes at 3.4 GHz for RHCP operation

narrower for higher operating frequencies. As shown in Fig. 3.44, the HPBWs of the proposed conformal antenna are 100°, 90° and 88° in the *xz*-plane at three frequencies for both operation modes, respectively. Similarly, the HPBWs are 98°, 90° and 84° in the *yz*-plane for different frequencies. Owning to the structural symmetry of the conformal element, the radiation patterns are basically the same for the LHCP and RHCP modes. Besides, the cross-polarization levels within HPBWs for both CP modes are below − 16.7, − 17.2 and − 16.4 dB at the frequency points of 2.6, 3 and 3.4 GHz, respectively. The front-to-back ratios (FBRs) of the radiation patterns are characterized to be more than 25, 17 and 15 dB at three frequencies for two CP operation modes, respectively.

3.2.2.2 Simulation and Measurement of the Conformal Wearable Array

To characterize the radiating performance of the hemispherical-conformal dual-CPs reconfigurable wearable array, a hemisphere-based conformal array prototype with a total size of $2\pi/3 \times 120^3$ mm^3 was fabricated and tested as described in Fig. 3.46. As shown in Fig. 3.46c and d, one 1:3 power divider together with three digital

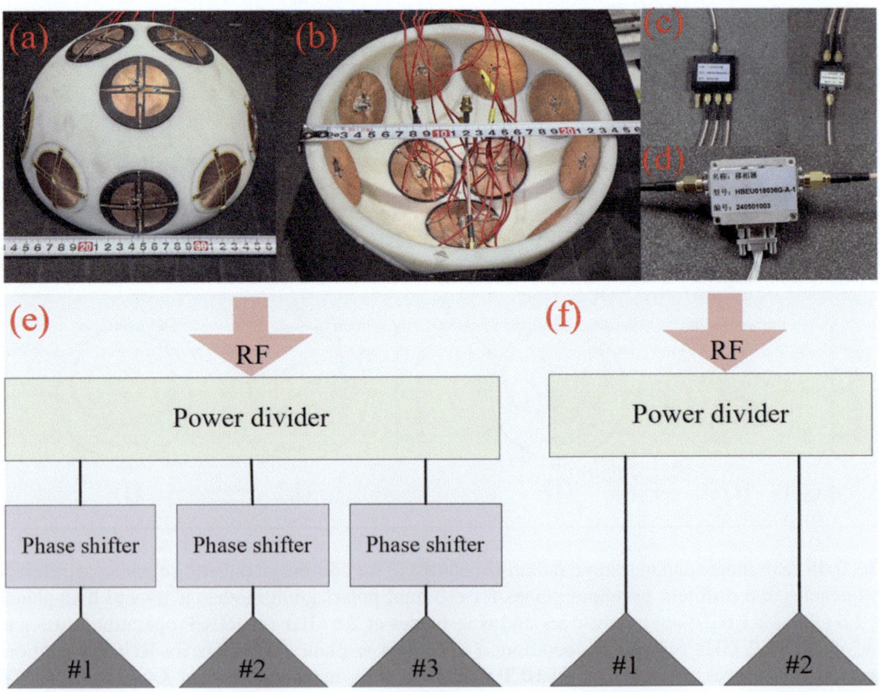

Fig. 3.46 Experimental details for characterizing the conformal wearable array: **a** external and **b** internal structure photographs of the fabricated conformal wearable array prototype; **c** the 1:2 and 1:3 power dividers and **d** the digital phase shifter; **e** the power-division phase-shifting network for the top three-element subarray; **f** the power-division network for the side bottom two-element subarray

phase shifters were utilized to provide phase compensation for the top three-element subarray, and one 1:2 power divider was used to feed the side bottom two-element subarray. Figure 3.46e and f illustrate the power division phase-shifting networks for the conformal subarrays. Table 3.5 lists the input phase combinations for different operation states of two conformal subarrays.

Figure 3.47 shows the measured reflection coefficients of the top three-element subarray and side bottom two-element subarray for the LHCP and RHCP operation modes, where both the cases of without and with human head were considered in the experiment. Those conformal subarrays were integrated with their corresponding power-division networks. Since the power dividers and phase shifters were purchased from a manufacturer without offering the simulated data, only the measured reflection coefficients for the overall conformal subarray systems are obtained and plotted. As shown in Fig. 3.47a and b, the two conformal subarrays yield ultrawide overlapped − 10-dB IMBWs of 85.7% (i.e., 1.8–4.5 GHz) and 69.8% (i.e., 2.03–4.21 GHz) for two CP operation modes, respectively. In the presence of a human head, it is seen from Fig. 3.47c and d that the measured overlapped IMBWs are 74.6% (i.e., 1.98–4.34 GHz) and 74.3% (i.e., 2.03–4.43 GHz) for the LHCP and RHCP modes, respectively. By comparing the measured results as shown in Fig. 3.47, it is found that the human head has very little influence on the available frequency range and operating bandwidth. It should be mentioned that, although there exist multiple resonant peaks in the curves as shown in Fig. 3.47c and d that are believed to result from the interference effect caused by the reflection of the human head, the amount of reflection is extremely small due to the high front-to-back ratio of the antenna, thus leading to the low S_{11} values over the operating frequency band. The measured reflection efficiency (i.e., $1 - |S_{11}|^2$) exceeds 90% over the operating frequency band for the case of either in the free space or with the human body. Finally, it is found that the measured IMBWs for those two conformal subarrays are much wider than those of the conformal radiating element, which may be caused by the insertional losses of the power-division phase-shifting networks.

In addition to the reflection coefficients, the radiation characteristics like realized gains and ARs of those two conformal subarrays were measured for the cases of without and with human head, respectively. For comparison, the simulation results

Table 3.5 Input phase combinations for different operation states of the conformal subarrays

Top three-element subarray 	LHCP	#1:0°	#2:120°	#3: – 120°
	RHCP	#1:0°	#2: – 120°	#3:120°
Side bottom two-element subarray 	LHCP	#1:0°	#2:0°	–
	RHCP	#1:0°	#2:0°	–

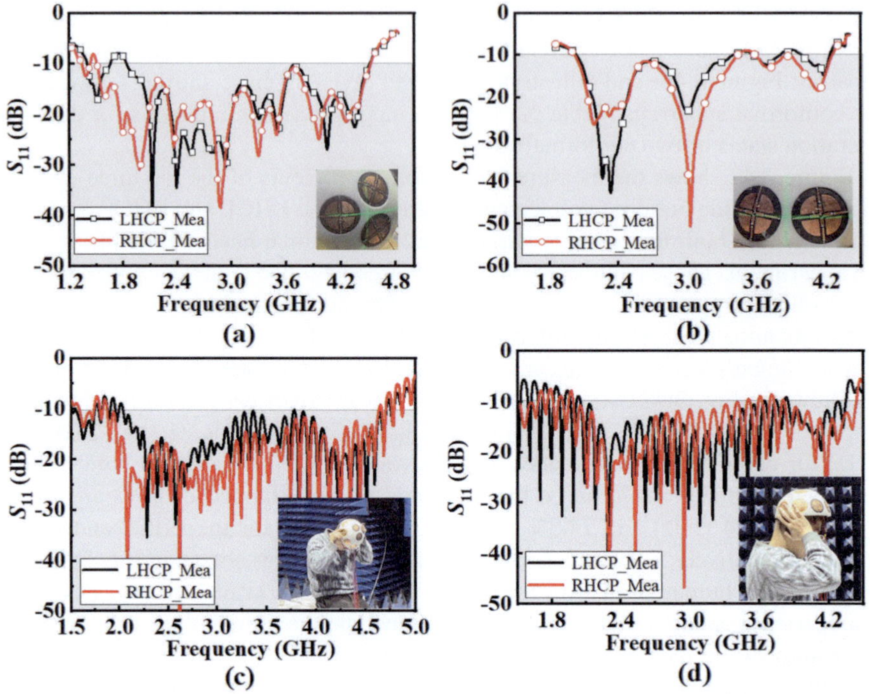

Fig. 3.47 Measured reflection coefficients for reconfigurable LHCP and RHCP modes of the: **a** top three-element subarray without human head; **b** side bottom two-element subarray without human head; **c** top three-element subarray with human head; **d** side bottom two-element subarray with human head

are also provided. As shown in Fig. 3.48a, a simplified human head model was established and integrated with the conformal array antenna for simulation analysis. As shown in Fig. 3.48b, the spherical head model with radius of 90 mm consists of six layers with actual dispersive dielectric properties for representing the skin, fat, cranial bones and cerebrospinal fluid, gray matter, and white matter, respectively [8]. Figure 3.49 shows the simulated and measured realized gains of the top three-element subarray and side bottom two-element subarray for two reconfigurable polarization modes. As illustrated in Fig. 3.49a, the measured peak realized gains of the three-element conformal subarray are 9.38 and 9.55 dBi for the LHCP and RHCP operation modes, respectively, which are slightly lower than the simulated results of 10.5 and 10.4 dBi. Moreover, the gain variation is less than 1.5 dB over the frequency range from 2.42 to 3.45 GHz. Seen from Fig. 3.49b, the measured peak gains for the two-element subarray are 8.1 and 8.15 dBi for both polarization modes, while the simulated results are 8.65 and 8.67 dBi. Besides, the gain variation is less than 1.5 dB over the frequency range of 2.2–3.5 GHz. For the case of the experiments with the human head, the measured peak realized gains of the three-element conformal subarray are 9.37 and 9.42 dBi for the LHCP and RHCP operation modes, as shown in

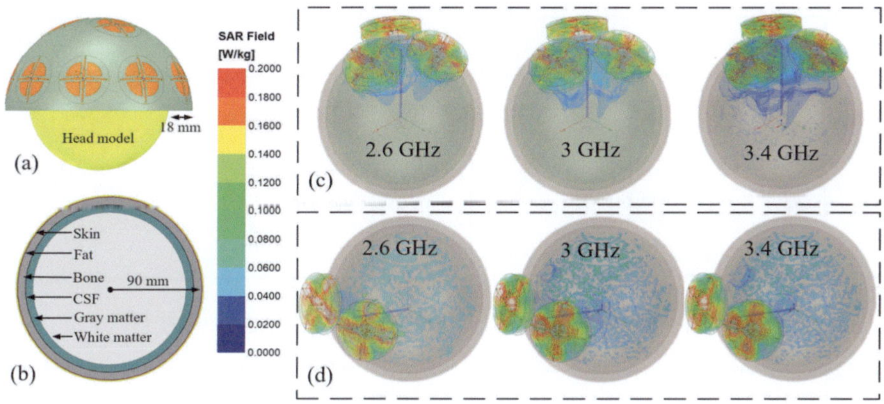

Fig. 3.48 **a** Structural geometry of the wearable array antenna with the human head model; **b** cross-sectional view of a spherical head model with six layers; **c** SAR distribution for the top three-element conformal subarray operating in the LHCP mode; **d** SAR distribution for the side bottom two-element conformal subarray operating in the LHCP mode

Fig. 3.49c. For the side bottom two-element subarray, it can be seen from Fig. 3.49d that the measured peak realized gains with the human head are 8.17 and 8.12 dBi for two operation modes. The effect of the human head on the radiation performance of the antenna is very little.

The measured gains are slightly lower than the simulated ones due to various experimental errors. Error analysis indicates that the discrepancies between the simulated and measured gains stem from multiple factors. Specifically, a ± 0.01 fluctuation in the loss tangent of the dielectric substrate causes a realized gain variation of ± 0.24 dB for the side bottom two-element subarray and ± 0.35 dB for the top three-element subarray, respectively. Besides, a ± 2.8° phase shift error introduced by the manufacturer offered phase shifter contributes a gain reduction up to 0.23 dB. During experiments, the non-ideal characteristics of the PIN diodes and manual soldering process for the SMA connectors can result in additional insertion losses. In addition, calibration errors for the standard-gain horn and the bending loss for the testing cables may also lead to reduced gain values. Finally, the radiation efficiency of the proposed conformal array is analyzed. Simulated results show that the peak radiation efficiencies of the top three-element subarray and the side bottom two-element array are 90% and 91% in the free space, respectively. Due to weak coupling between the antenna and human tissue, the radiation efficiencies decrease by approximately 8% when those two subarrays are worn on the human head.

Figure 3.50 shows the simulated and measured ARs of the top three-element subarray and side bottom two-element subarray for two reconfigurable polarization modes. As illustrated in Fig. 3.50a, an ultrawide overlapped 3-dB ARBWs more than 2–4 GHz was measured in free space for LHCP and RHCP operation modes, which agrees reasonably well with the analysis results as detailed in Section II-B. Compared to simulation, a little bit higher values for the measured ARs are due to the

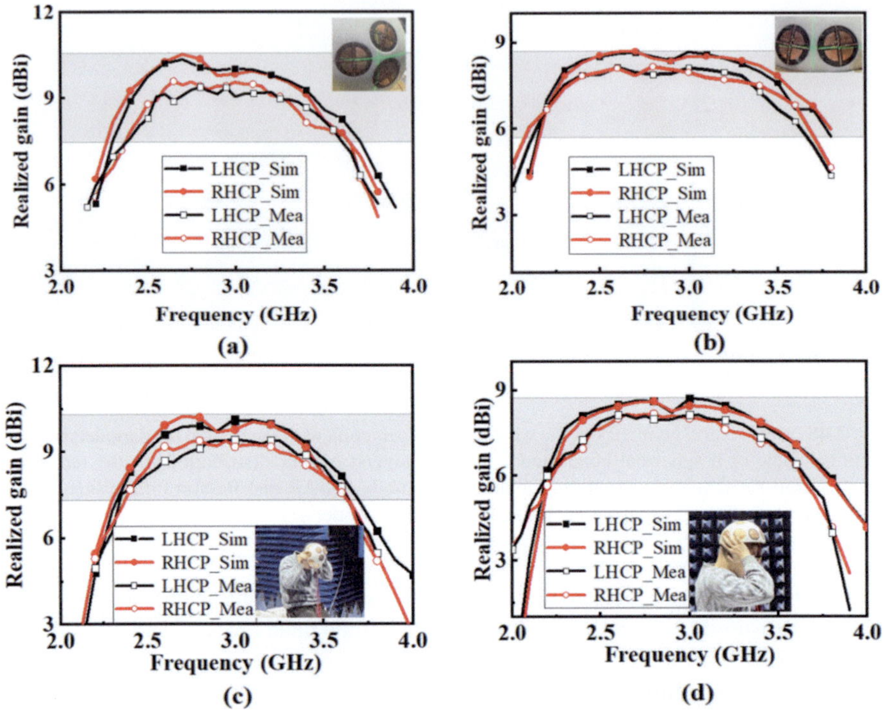

Fig. 3.49 Simulated and measured realized gains for reconfigurable LHCP and RHCP modes of the: **a** top three-element subarray without human head; **b** side bottom two-element subarray without human head; **c** top three-element subarray with human head; **d** side bottom two-element subarray with human head

assembly errors. Figure 3.50b shows that the measured overlapped 3-dB ARBW is 29.6% (i.e., 2.56–3.45 GHz) for the two-element subarray for two reconfigurable CP modes, where the frequency range moves upward by around 40 MHz as compared with the simulated result due to fabrication errors. As shown in Fig. 3.50c and d, the measured overlapped 3-dB ARBWs for two CP modes with the human head also cover the band of 2–4 GHz for the top three-element subarray and achieve 29.3% (i.e., 2.53–3.4 GHz) for the side bottom two-element subarray, respectively. As shown in Figs. 3.49 and 3.50, after calibrating the insertion losses introduced by the power-division phase-shifting networks, the experimental and simulation results still show some discrepancy. The minor differences may be caused by the errors in protype fabrication and assembly as well as feeding current magnitudes and phases for the power-division phase-shifting networks.

Figure 3.51 shows the simulated and measured normalized radiation patterns of the top three-element conformal subarray at 2.6, 3 and 3.4 GHz for two reconfigurable CP operation modes. It is clearly seen that the measured HPBWs of the proposed array antenna in the *xz*-plane are 57°, 48° and 47° at three frequencies for the LHCP and RHCP operation modes, respectively. Similarly, the HPBWs in the *yz*-plane were

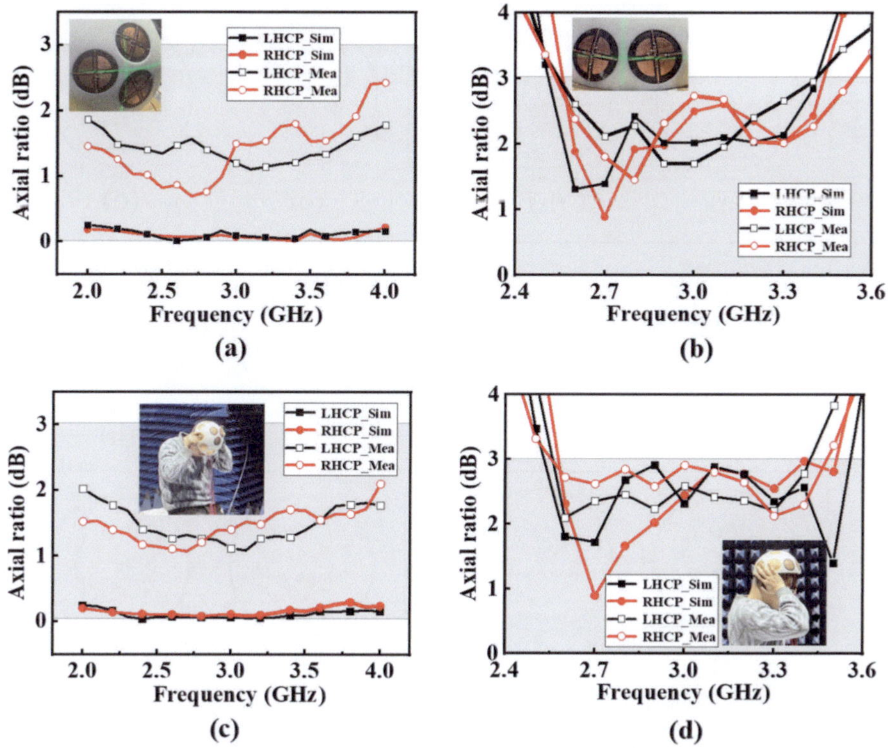

Fig. 3.50 Simulated and measured ARs for reconfigurable LHCP and RHCP modes of the: **a** top three-element subarray without human head; **b** side bottom two-element subarray without human head; **c** top three-element subarray with human head; **d** side bottom two-element subarray with human head

measured to be 56°, 47° and 46°, respectively. Obviously, the radiation patterns for the LHCP and RHCP modes are almost the same. Besides, the cross-polarization levels within HPBWs are below − 15.9, − 20.1 and − 14 dB at the frequencies of 2.6, 3 and 3.4 GHz for two CP modes, respectively. The key contributing factor lies in the rotationally symmetric configuration of the top three-element subarray (i.e., *Case 1* as shown in Figs. 3.33, 3.34 and 3.35), which helps to significantly suppress the cross-polarization components through coherent superposition. This configuration effectively enhances the polarization purity over the entire operating bandwidth, for which the underlying mechanisms are thoroughly discussed in Sect. 3.1.2.1. The measured FBRs of the radiation patterns was found to be higher than 12.9, 12.2 and 10 dB at three frequencies for both polarization modes. Overall, the measured and simulated radiation patterns are in good agreement.

Figure 3.52 shows the simulated and measured normalized radiation patterns of the side bottom two-element conformal subarray at 2.6, 3 and 3.4 GHz for two CP operation modes. It is clearly seen that the measured HPBWs of the proposed array antenna in the $x''z''$-plane are 94°, 95° and 93° for both operation modes at three

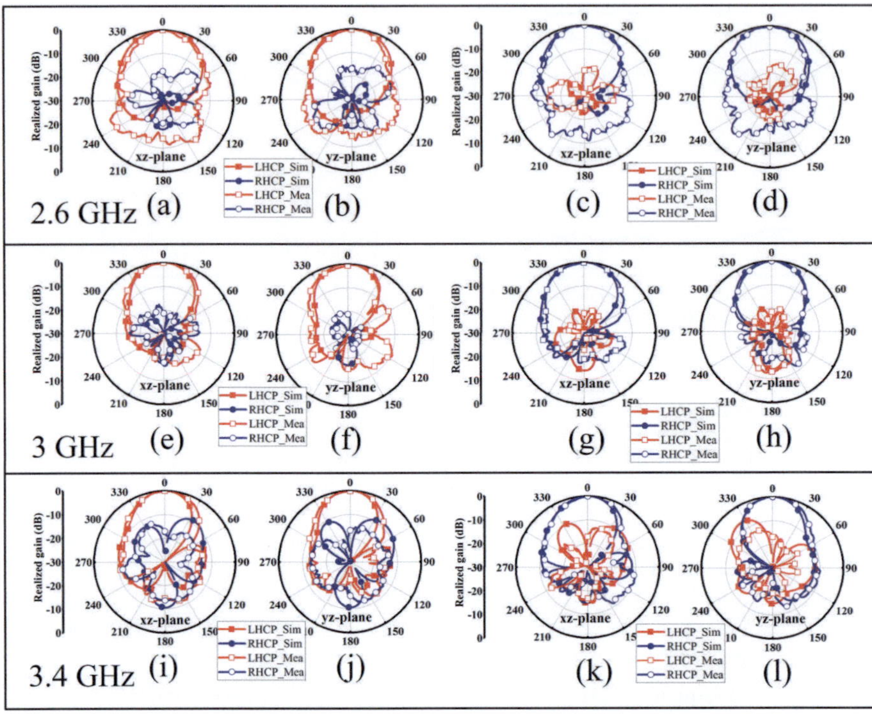

Fig. 3.51 Simulated and measured radiation patterns of the top three-element conformal subarray at different frequencies and different principal planes for different polarization modes: **a** *xz*- and **b** *yz*-planes at 2.6 GHz for LHCP operation; **c** *xz*- and **d** *yz*-planes at 2.6 GHz for RHCP operation; **e** *xz*- and **f** *yz*-planes at 3 GHz for LHCP operation; **g** *xz*- and **h** *yz*-planes at 3 GHz for RHCP operation; **i** *xz*- and **j** *yz*-planes at 3.4 GHz for LHCP operation; **k** *xz*- and **l** *yz*-planes at 3.4 GHz for RHCP operation

frequencies, respectively. Similarly, the HPBWs in the $y''z''$-plane were measured to be 43°, 41° and 39°, respectively. Besides, the cross-polarization levels within HPBWs are below − 14.1, − 12.2 and − 12.5 dB at the frequencies of 2.6, 3 and 3.4 GHz for both CP modes, respectively. The measured FBRs of the radiation patterns are all higher than 15.3, 11.6 and 13.1 dB at three frequencies for two CP operation modes, respectively. Hence, all measured and simulated radiation patterns are in very good agreement. Owing to the inconvenience of rotating the human body, the radiation patterns of the antenna were not measured for the case of with the human head.

To verify that the proposed hemispherical conformal array antenna has the capablity of 110° × 360° beam coverage, we combine the measured radiation patterns of the top three-element subarray and side bottom two-element subarray in the xz-plane, which are plotted in Fig. 3.53. Clearly, the combined radiation patterns shows that the proposed conformal antenna exhibits exhibit a beam coverage of 110° in the elevational plane at the frequencies of 2.6, 3.0, and 3.4 GHz. On the other hand, by

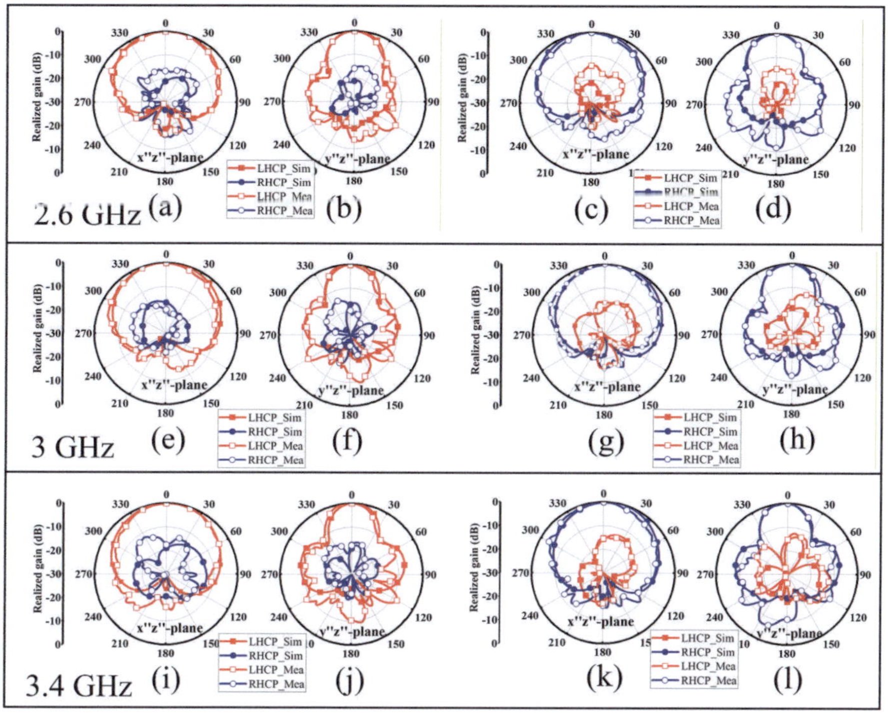

Fig. 3.52 Simulated and measured radiation patterns of the side bottom two-element conformal subarray at different frequencies and different principal planes for different polarization modes: **a** $x''z''$- and **b** $y''z''$-planes at 2.6 GHz for LHCP operation; **c** $x''z''$- and **d** $y''z''$-planes at 2.6 GHz for RHCP operation; **e** $x''z''$- and **f** $y''z''$-planes at 3 GHz for LHCP operation; **g** $x''z''$- and **h** $y''z''$-planes at 3 GHz for RHCP operation; **i** $x''z''$- and **j** $y''z''$-planes at 3.4 GHz for LHCP operation; **k** $x''z''$- and **l** $y''z''$-planes at 3.4 GHz for RHCP operation

sequentially exciting the side bottom two-element subarrays at different orientations, the combined radiation patterns can be sequentially rotated around the z-axis thus achieving a 360° full-azimuth coverage. To avoid excessive length of the article, the combined radiation patterns showing 360° coverage in the azimuthal plane are not shown for brevity. Therefore, a large coverage range of 110° × 360° can be well implemented via flexible beam switching between the top three-element subarray and sequentially excited side bottom two-element subarrays.

Considering the conformal array antenna is proposed for wearable applications, the specific absorption rate (SAR) that evaluates the amount of electromagnetic radiation leaking into the head tissue is a key performance indicator. According to the IEEE C95.1-1999 and C95.1-2005 standards [9], the human health would not be threatened if the SAR is limited to 1.6 W/kg averaged over 1 g tissue. Figure 3.6c and d illustrate the simulated SARs for two conformal subarrays operating at 2.6, 3 and 3.4 GHz for the LHCP mode. Assuming that the typical input power of a single conformal element is 100 mW, the maximum input power of the proposed conformal

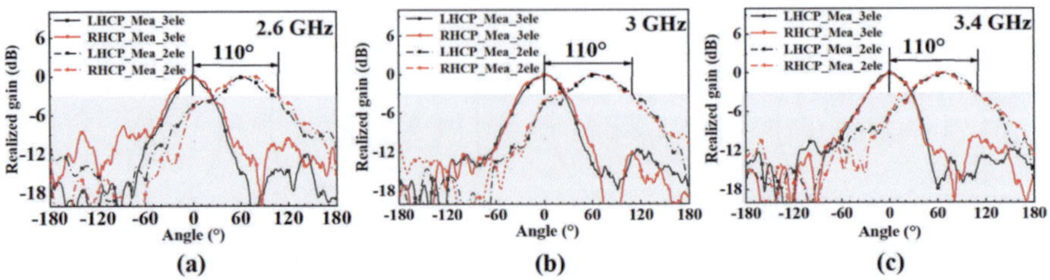

Fig. 3.53 Combination of the measured radiation patterns of the hemispherical-conformal subarrays in the xz-plane at different frequencies for different polarization modes: **a** 2.6 GHz for LHCP and RHCP operation; **b** 3 GHz for LHCP and RHCP operation; **c** 3.4 GHz for LHCP and RHCP operation

antenna is 300 mW. Considering the power losses introduced by the phase shifters and power dividers, the minimum total input power required for normal operation of the proposed antenna is 457 mW. Simulation results indicate that the SAR value is below 0.2 W/kg, which is well below the safety standard limit. Thus, the proposed helmet-based conformal array antenna can be safely worn on the human head.

References

1. X.-X. Yang, B.-C. Shao, F. Yang, A. Z. Elsherbeni, and B. Gong, "A polarization reconfigurable patch antenna with loop slots on the ground plane," *IEEE Antennas Wireless Propag. Lett.*, vol. 11, pp. 69-72, Mar. 2012.
2. F. Wu and K. M. Luk, "Single-port reconfigurable magneto-electric dipole antenna with quad-polarization diversity," *IEEE Trans. Antennas Propag.*, vol. 65, no. 5, pp. 2289–2296, May. 2017
3. M.-C. Tang, Z. Wu, T. Shi, and R. W. Ziolkowski, "Electrically small, low-profile, planar, huygens dipole antenna with quad-polarization diversity," *IEEE Trans. Antennas Propag.*, vol. 66, no. 12, pp. 6772-6780, Dec. 2018.
4. H. T. Hui, "A practical approach to compensate for the mutual coupling effect in an adaptive dipole array," *IEEE Transactions on Antennas and Propagation*, vol. 52, no. 5, pp. 1262-1269, May 2004.
5. N. Saha, E. P. Alvarez and I. Mahbub, "A novel theoretical modeling of the received power for phased array-based wireless power transfer system in the near-field region," *IEEE Open Journal of Antennas and Propagation.*, vol. 5, no. 6, pp. 1476–1488, Dec. 2022.
6. T. Zhang and W. Ser, "Robust beampattern synthesis for antenna arrays with mutual coupling effect," *IEEE Transactions on Antennas and Propagation*, vol. 59, no. 8, pp. 2889-2895, Aug. 2011.
7. B. D. Braaten, S. Roy, I. Irfanullah, S. Nariyal, and D. E. Anagnostou, "Phase-compensated conformal antennas for changing spherical sur faces," *IEEE Trans. Antennas Propag.*, vol. 62, no. 4, pp. 1880–1887, Apr. 2014.
8. A. Hossain, M. T. Islam, M. E. H. Chowdhury, and M. Samsuzzaman, "A grounded coplanar waveguide-based slotted inverted delta-shaped wideband antenna for microwave head imaging," IEEE Access, vol. 8, pp. 185698–185724, 2020.
9. A. Alemaryeen and S. Noghanian, "On-body low-profile textile antennawith artificial magnetic conductor," *IEEE Trans. Antennas Propag.*, vol. 67, no. 6, pp. 3649–3656, Jun. 2019.

Chapter 4
DOA Estimation and Beamforming Algorithms Based on Conformal Sparse Arrays

Conformal array antennas are extensively used in communication and radar fields due to their ability to adapt to carrier contours and excellent aerodynamic properties. However, increasing the size of such arrays significantly raises the number of array elements, severely impacting system cost and complexity while creating an urgent need for sparse optimization. To date, research on conformal sparse arrays remains relatively limited and insufficiently explored, facing challenges including complex design, high costs, and computationally intensive algorithms. Therefore, this chapter focuses on conformal sparse arrays, proposing a coherent signal DOA estimation algorithm based on conformal circular arrays, a sparse optimization algorithm for conformal arrays, and an interference suppression beamforming algorithm for conformal sparse arrays.

4.1 Coherent Signal DOA Estimation Algorithm Based on Conformal Circular Arrays

In modern wireless communication applications, there are complex multipath effects and coherent signals which pose significant challenges to DOA estimation. When there are coherent signals among the multiple received signals, the received covariance matrix of the array ceases to be full-rank. This results in the loss of useful signal information, causing a mismatch between the estimated DOA and the actual DOA, and consequently leading to the failure of subspace-based DOA estimation algorithms. A uniform circular array is an antenna array with its elements arranged in a specific circular configuration. Its elements are uniformly distributed along a circular (or approximately circular) aperture, exhibiting azimuthal performance invariance. This enables more flexible reception of signals from arbitrary azimuthal directions, offering distinct application advantages over uniform linear arrays. Therefore, this study investigates coherent signal DOA estimation algorithms based on circular

X. Gao et al., *Conformal Array Antennas and Beamforming Technologies for Emergency Communication Applications*, Modern Antenna,
https://doi.org/10.1007/978-981-95-2905-6_4

arrays. It proposes several effective coherent signal DOA estimation algorithms to address the limitations of subspace-based algorithms, leveraging the inherent properties of circular arrays to achieve omnidirectional scanning in azimuth direction.

In this section, based on the L1-SVD algorithm, we propose one-dimensional coherent signal DOA estimation algorithms, which are tailored for uniform circular arrays and sparse circular arrays separately. Targeting the rank reduction problem of covariance matrix of coherent signals, this approach transforms coherent signal DOA estimation into an optimization problem that bypasses covariance matrix calculation. It simultaneously leverages the inherent wide-angle resolution capability of uniform circular arrays in the azimuth plane to mitigate the limitations of subspace-based algorithms. To address the computational intensity and estimation challenges of two-dimensional coherent signal DOA estimation, this section exploits the unique spatial characteristics of circular arrays. We propose a two-dimensional coherent signal DOA estimation algorithm based on conformal circular arrays. This algorithm reduces the computational load associated with covariance matrix processing, enables simultaneous estimation of both azimuth and pitch angles, and achieves DOA resolution for signals exhibiting distinct correlation characteristics.

4.1.1 One-Dimensional Coherent Signal DOA Estimation Algorithm of Conformal Circular Arrays Based on Sparse Reconstruction

The application of sparse reconstruction algorithm in DOA estimation needs to regard the source as a point source, and the number of sources is very small compared with the number of array elements, so the DOA of the source can be considered as sparse in the array space. The source in the whole spatial domain is sparse relative to the space where the array is located, and the signal to be estimated also has the natural spatial sparsity in the whole angle space. Combined with the compressive sensing theory [1], the DOA estimation problem can be converted into a convex optimization second order cone programming (SOCP) to solve with a small number of spatial points. This approach efficiently yields the DOA information for the target signals.

Under the support of sparse reconstruction theory, the DOA estimation problem can be converted into the sparse reconstruction problem of spatial spectrum by constructing a complete sparse representation basis by a specific spatial grid division, and the sparse representation of the signal of the receiving array can be obtained by using the optimization algorithm. The sparse reconstruction algorithm divides the whole space to be measured into uniform grids, and each grid corresponds to an angle value, as shown in Fig. 4.1. If a signal is detected in the range of a grid, the grid corresponds to a non-zero value, and if there is no signal in the range of the grid, the position is zero. After this process, the DOA estimation problem is transformed into solving the position of the grid non-zero value under sparse reconstruction. After

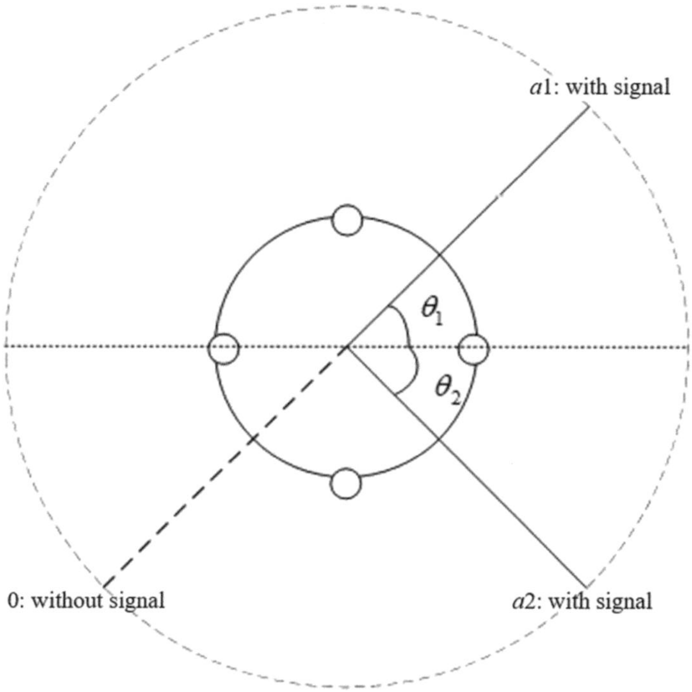

Fig. 4.1 Schematic of uniform circular array sparse mesh partitioning

finding the corresponding position and obtaining the angle value, the DOA of the signal to be estimated can be obtained.

4.1.1.1 DOA Estimation Algorithm Based on Sparse Reconstruction

The L1-SVD algorithm is the most representative algorithm in convex optimization tools for solving sparse reconstruction problems and also belongs to the category of grid partitioning type sparse reconstruction algorithms. The key of solving this algorithm is: By performing singular value decomposition on the covariance matrix of the received signal, the high-dimensional received signal model measured by a large number of snapshots is transformed into a low-dimensional signal model measured by the number of sources to be estimated. After the transformation, common nonlinear optimization algorithms can be used to solve the problem. The complete steps of the L1-SVD algorithm are as follows.

Firstly, constructing a complete spatial sparse representation basis enables the received array signal to be sparsely represented. The DOA estimation algorithm divides the spatial grid by using the method of equal-angle separated sampling. The array receiving model represented by the spatial sparse basis is expressed as Formula (4.1)

$$Y = AX + N \tag{4.1}$$

where $Y \in \mathbb{C}^{M \times T}$ is the output signal after sparse representation, $A \in \mathbb{C}^{M \times N_\theta}$ is the direction vector basis of sparse representation, $N \in \mathbb{C}^{M \times T}$ is the noise sampling signal, M is the number of elements of the uniform array, T is the number of time-domain snapshots, and N_θ is the number of spatial grids at equal angles.

Due to the interference caused by noise superposition, additional non-zero terms containing noise signals appear in the sparse representation basis of signal X. In order to recover the sparse representation vector of the signal from the noise signal superposition, minimizing the l_0 norm $\|X\|_0$ needs to be taken as the objective function, that is, to minimize the number of non-zero terms in the signal

$$\|X\|_0 = \text{card}\left(\left\{(i,j)|X_{ij} \neq 0\right\}\right) \tag{4.2}$$

However, the l_0 norm is non-convex and difficult to solve. Therefore, in practical applications, it is generally convex relaxation and transformed into the l_1 norm $\|X\|_1$ for processing, that is, minimizing the sum of the absolute values of the elements in the solution signal vector

$$\|X\|_1 = \sum |X_{ij}| \tag{4.3}$$

Furthermore, we often use the "Frobenius" norm of the observation matrix to measure the gap between the observation matrix Y and the ideal receiving matrix AX

$$\|X\|_F = \sqrt{\sum_i \sum_j |x_{ij}|^2} \tag{4.4}$$

$$\|Y - AX\|_F^2 \leq \beta^2 \tag{4.5}$$

where β^2 is generally the power of the noise signal.

The above optimization problems can also be written as the unconstrained form, that is, $\min ||Y - A||_F^2 + \lambda||X||_1$, where λ is used to control the sparsity of the spatial spectrum and the balance of the norm offset scale, with a range of approximately 0.1–10.

Singular value decomposition can analyze the main components of a matrix. In DOA estimation, the characteristics of singular value decomposition are very suitable for analyzing the signal components disturbed by noise. Its decomposition process is as follows.

Firstly, perform singular value decomposition on the received signal matrix Y

$$Y = ULV^H \tag{4.6}$$

where the matrix $U \in \mathbb{C}^{M \times M}$, the matrix $V \in \mathbb{C}^{T \times T}$, and the matrix $L \in \mathbb{C}^{M \times T}$, where L is composed of a diagonal matrix of order $M \times M$ and a null matrix of order $M \times (T - M)$, and the diagonal matrix contains K large elements and $M - K$ small elements, where K is the number of sources to be estimated. In a diagonal matrix, there is a difference of order of magnitude between large elements and small elements, and these large elements in the diagonal matrix can be easily filtered out.

Define a choice matrix $D_K \in \mathbb{C}^{T \times K}$ consisting of a $K \times K$ identity matrix and a $(T - K) \times K$ null matrix, and the choice matrix has the following properties

$$Y_{SV} = YVD_K \tag{4.7}$$

$$Y_{SV} = ULD_K \tag{4.8}$$

Then, a new receiving matrix $Y_{SV} \in \mathbb{C}^{M \times K}$ is obtained, which reduces the dimension of the signal, making the algorithm faster and more efficient, and more suitable for the condition of large number of snapshots. After linear transformation, the receiving model of the signal becomes

$$Y_{SV} = AS_{SV} + N_{SV} \tag{4.9}$$

Correspondingly, the l_1 norm optimization problem has transformed into solving the expreseesion, denoted as $\min\limits_{S_{SV}} ||Y_{SV} - AS_{SV}||_F^2 + \lambda||S_{SV}||_1$.

Finally, solve this nonlinear problem to obtain the DOA of the signal to be estimated.

4.1.1.2 One-Dimensional Coherent Signal DOA Estimation Algorithm and Simulation Results of Uniform Circular Arrays

Based on the L1-SVD algorithm and designed with the idea of SOCP, the l_1 norm optimization problem can be further simplified and solved.

The definition of SOCP is expressed as

$$\mathcal{C}_k = \left\{ \begin{bmatrix} u \\ t \end{bmatrix} : u \in \mathbb{R}^{K-1}, t \in \mathbb{R}, \|u\| \le t \right\} \tag{4.10}$$

The constraints of SOCP are expressed as

$$\|Ax + b\| \le c^T x + d \tag{4.11}$$

that is

$$\begin{bmatrix} A \\ c^T \end{bmatrix} x + \begin{bmatrix} b \\ d \end{bmatrix} \in \mathcal{C}_k \tag{4.12}$$

where A is the direction vector of the uniform circular array. The nonlinear optimization problem in the previous subsection can be transformed into a SOCP problem, that is, $\min_{p,q,r,S_{SV}} (p + \lambda q)$, where the constraint condition, described as Formula (4.13), is that the norm of the affine transformation of the variable to be optimized is no more than another affine transformation.

$$\begin{gathered} vec\{Y_{SV} - AS_{SV}\}_2^2 \leq p \\ I^T r \leq q \\ S_{SV}(i, :)_2 \leq r_i \end{gathered} \tag{4.13}$$

Therefore, the variable correspondence of the SOCP is

$$x = \begin{bmatrix} p \\ q \\ r \\ vec\{S_{SV}\} \end{bmatrix} \tag{4.14}$$

The first constraint condition is transformed into

$$vec\{Y_{SV}\} - (I \otimes A)vec\{S_{SV}\} \leq p \tag{4.15}$$

which conforms to the definition of the constraint conditions of SOCP.

Figure 4.2 shows the comparison of the one-dimensional DOA estimation effects of the proposed algorithm and the MUSIC algorithm on a uniform circular array. The simulation conditions are as follows: the number of array elements of the uniform circular array is 25, the signal wavelength is λ, the distance between proximity array elements is $\lambda/2$, the incident signals are two incoherent sources to be estimated, the incident directions are 100° and 120° respectively, and the SNR is set to 20 dB. On the premise of realizing the DOA estimation of the signal, the nulls formed by the algorithm in this study is below − 100 dB, which is much lower than − 46 dB of the MUSIC algorithm. The performance of the algorithm is far superior to that of the MUSIC algorithm, which is of great help for achieving efficient multi-user detection and interference management.

Figure 4.3 presents a comparison of the resolution of the proposed algorithm and the MUSIC algorithm for proximity angle sources. Other conditions remain unchanged, but the incident directions change to 100° and 103°. The nulls formed by the proposed algorithm during proximity angle resolution is below − 40 dB, which is much lower than − 10 dB of the MUSIC algorithm. Therefore, the proximity angle resolution is superior to that of the MUSIC algorithm. The proximity angle resolution of the proposed algorithm is about 3°, which can identify and locate the target with high resolution and improve the detection ability of the array system.

Figure 4.4 presents the ability of the proposed algorithm to distinguish the number of coherent sources. The number of array elements is 25, the signal wavelength is λ, and the spacing between proximity array elements is $\lambda/2$. Under the condition of a

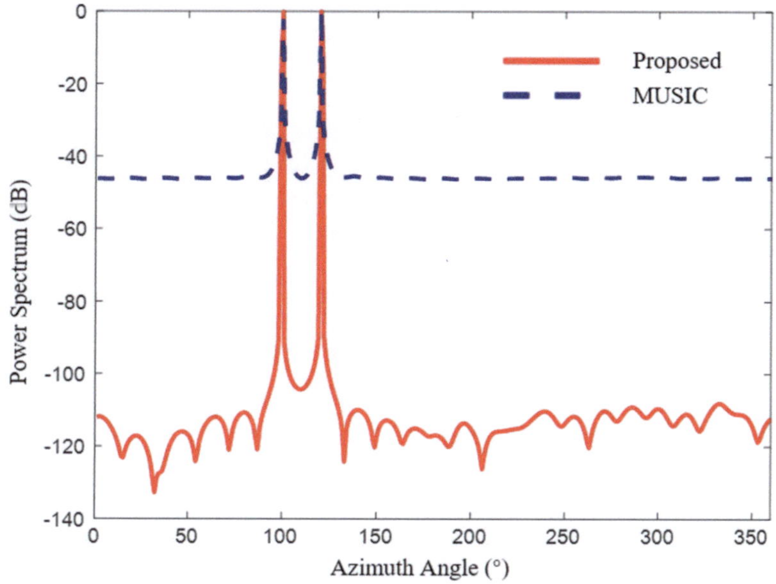

Fig. 4.2 The comparison of the proposed algorithm and the MUSIC algorithm in estimating nulls

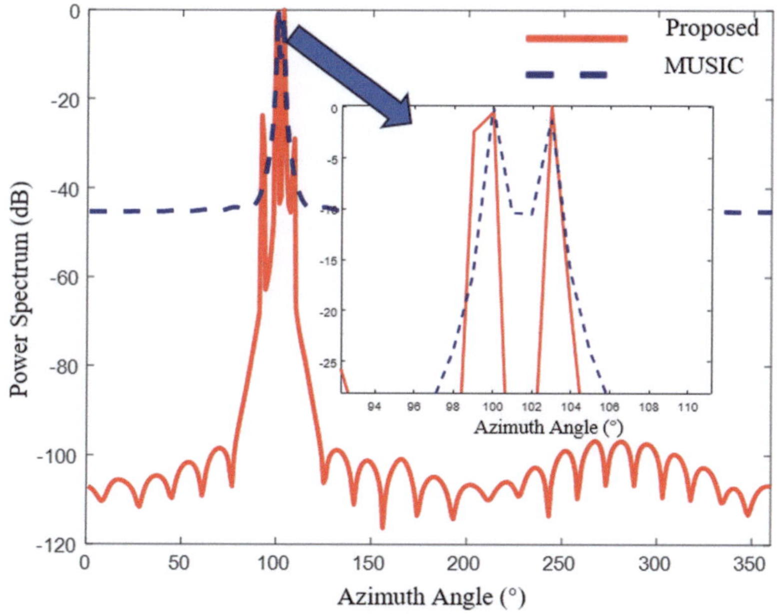

Fig. 4.3 The comparison of the proposed algorithm and the MUSIC algorithm in estimating proximity angles

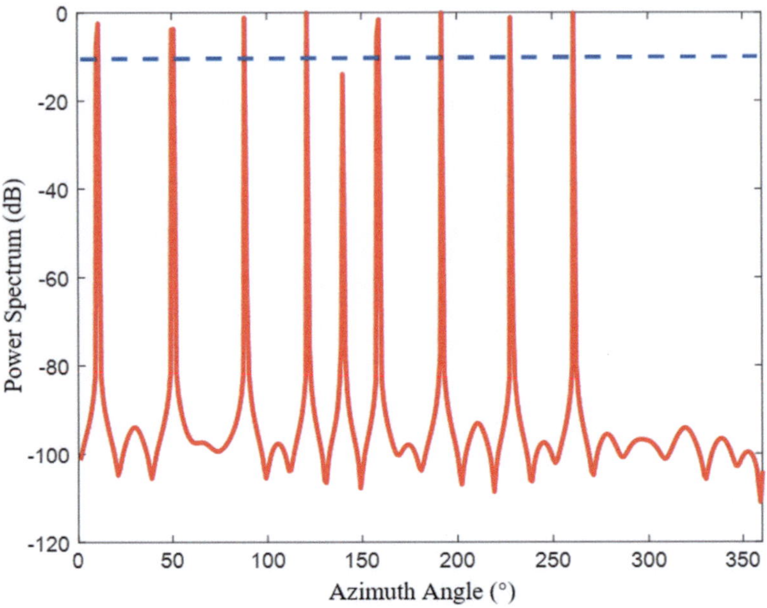

Fig. 4.4 Schematic of the maximum estimated number of sources

SNR of 20 dB, a total of 8 coherent signals are incident on the receiving array, and the incident directions are 10°, 50°, 90°, 120°, 160°, 190°, 230°, and 260° respectively. The simulation results show that this algorithm can distinguish up to 8 directions of arrival at most, and the beam of the algorithm has the uniform and symmetrical characteristics of a circular array, which can achieve 360° estimation characteristics in the azimuth direction. Figure 4.5 shows the state of the algorithm when there are incident failures of 9 coherent sources. Although a wave crest can be formed in the direction of the incoming wave, the wave crest at such a 90° is statistically indistinguishable.

4.1.1.3 One-Dimensional Coherent Signal DOA Estimation Algorithm and Simulation Results of Sparse Circular Arrays

The number of elements in a conformal antenna array increases with the increase of array size. Too many elements will increase the cost and complexity of the antenna system. In order to reduce the cost of the antenna system and avoid size limitations, this study proposes an algorithm for DOA estimation using a sparse circular array, which is more suitable for some resource-constrained or space-constrained application scenarios such as emergency rescue environments.

According to the compressive sensing theory, for signals with spatial sparsity, sampling can be carried out with a low number of sampling points, and optimization

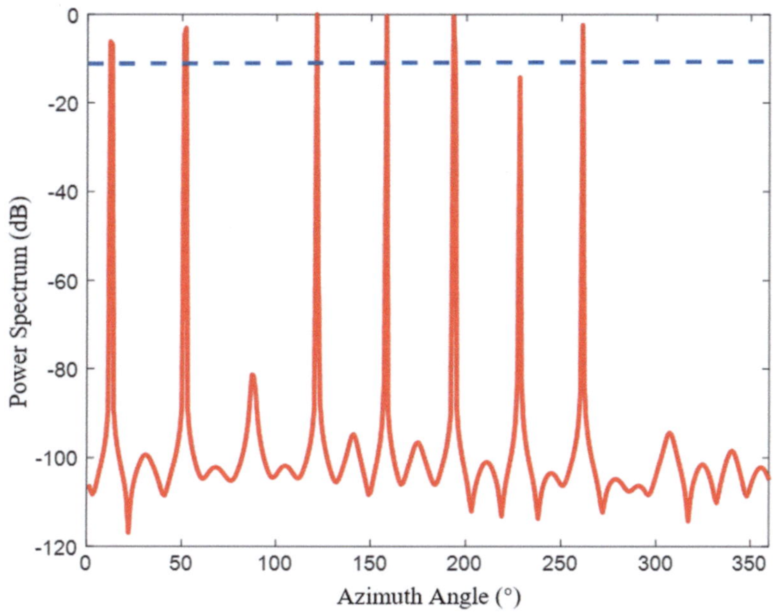

Fig. 4.5 Schematic of source estimation failure

algorithms can be used to solve the problem. Therefore, the sparse reconstruction algorithm can naturally be applied to the field of DOA estimation algorithms for sparse arrays. Sparse arrays reduce the number of some antenna array elements and sacrifice certain DOA estimation performance while ensuring the accuracy requirements of DOA estimation. This can save the cost and resources of the antenna system and improve the flexibility of the system.

The signal reception model of the sparse circular array can be expressed as

$$X_{array}(l) = As(l)G_{array} + n(l) \tag{4.16}$$

where $X_{array}(l)$ is the receiving signal matrix of sparse array, A is the spatial direction vector matrix, $s(l)$ is the spatial signal vector matrix, G_{array} is the array elements layout of the sparse circular array, and $n(l)$ is the noise vector matrix.

Similarly, the sparse basis matrix of the L1-SVD algorithm for sparse circular arrays can be expressed as

$$A_{array} = AG_{array} \tag{4.17}$$

Then its sparse signal receiving matrix is

$$Y_S = A_{array}X_{array} + N \tag{4.18}$$

Similarly, perform singular value decomposition on the sparse receiving matrix

$$Y_{SSV} = Y_S V D_K \tag{4.19}$$

$$Y_{SSV} = ULD_K \tag{4.20}$$

A new sparse receiving matrix is obtained and represented in the following way

$$Y_{SSV} = A_{array} S_{SV} + N_{SV} \tag{4.21}$$

Correspondingly, the l_1 optimization problem is expressed as $\min\limits_{S_{SV}}(Y_{SSV} - AS_{SVF}^2 + \lambda \tilde{s}_1^{l_2})$.

Solve the above problems through SOCP, that is, $\min\limits_{p,q,r,S_{SV}} \quad p + \lambda q$, where the constraint condition, described as Formula (4.22), is that the norm of the affine transformation of the variable to be optimized is no more than another affine transformation.

$$\begin{gathered} vec\{Y_{SSV} - A_{array} S_{SV}\}_2^2 \le p \\ I^T r \le q \\ S_{SV}(i, :)_2 \le r_i \end{gathered} \tag{4.22}$$

$$x = \begin{bmatrix} p \\ q \\ r \\ vec\{S_{SV}\} \end{bmatrix} \tag{4.23}$$

$$vec\{Y_{SSV}\} - (I \otimes A_{array}) vec\{S_{SV}\} \le p \tag{4.24}$$

The one-dimensional DOA effect of the proposed algorithm on the sparse circular array is discussed below. The simulation conditions are as follows: the number of elements in the non-sparse antenna array is 25, the spacing between proximity elements is $\lambda/2$, the SNR is 20 dB, the sparsity is 44%, and the number of elements in the sparse array is 12. Its sparse layout is shown in Fig. 4.6.

Figure 4.7 shows the comparison of estimation effect between sparse circular array and non-sparse circular array. The DOA angle estimation with sparse array is about 10 dB higher than that with non-sparse array. The estimation performance of sparse array is slightly lower than that of full array, but it is still -50 dB lower than that of MUSIC algorithm. Under the premise of maintaining a certain estimation effect, sparse array reduces the number of array elements, reduces the system power consumption, burden and implementation cost, and has important application value for low-power systems such as UAVs.

Figure 4.8 shows the proximity angle resolution of the proposed algorithm on the sparse circular array. Other conditions remain unchanged, the incident angle changes

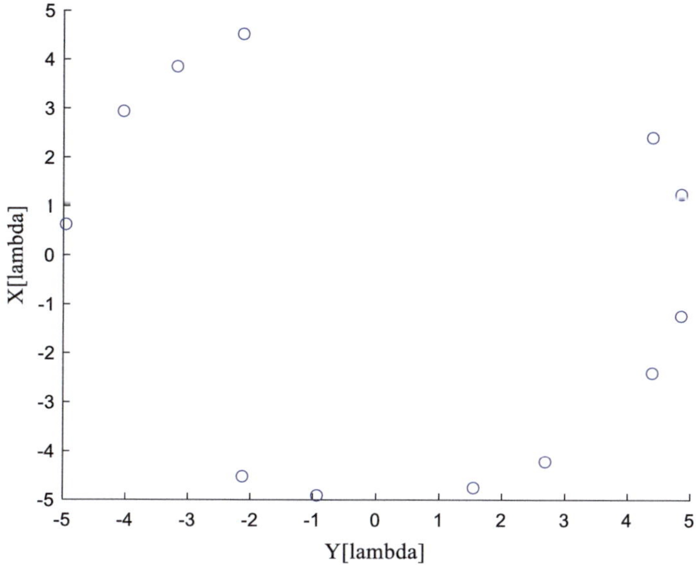

Fig. 4.6 Schematic of the layout of sparse circular array

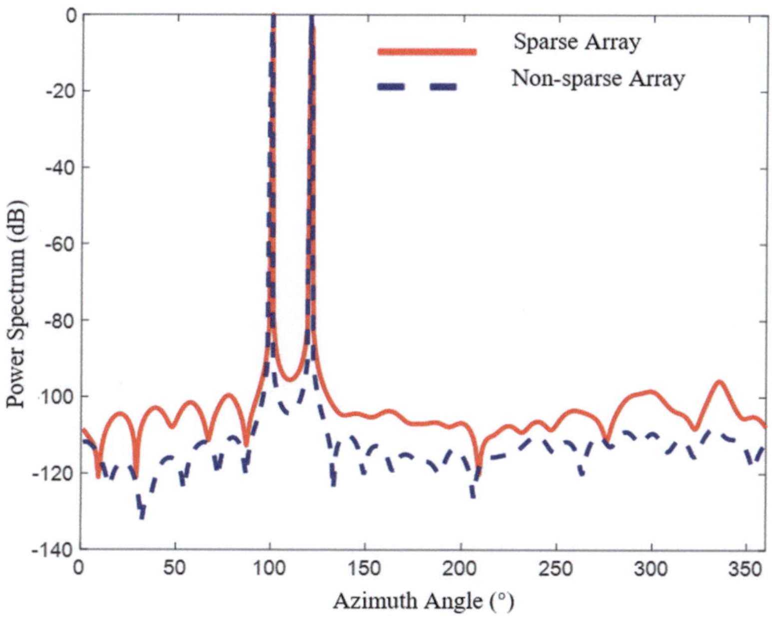

Fig. 4.7 Comparison of estimation effects between sparse and non-sparse arrays

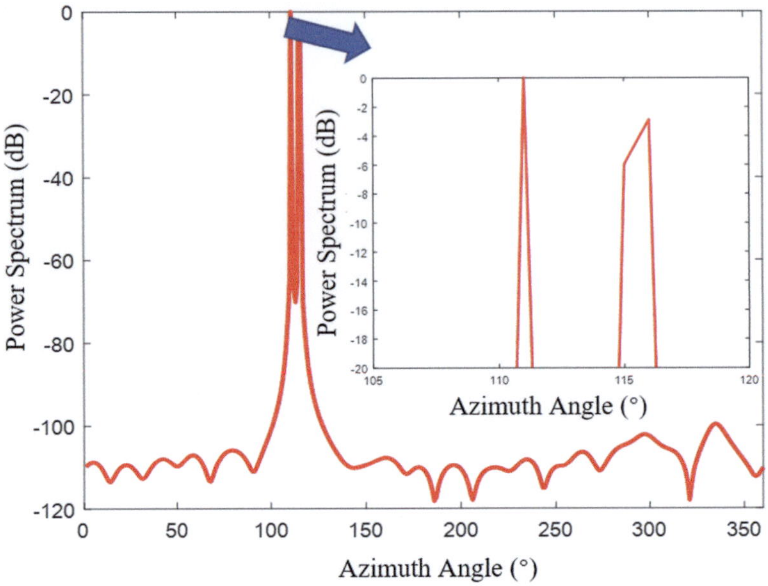

Fig. 4.8 Proximity angle resolution for sparse arrays

to 111° and 116°, and the maximum proximity angle resolution is about 5°, which is lower than that of the non-sparse array, but its performance is still better than that of the MUSIC algorithm.

Figure 4.9 shows the maximum number of source resolution capability of the sparse circular array. The simulation conditions are as follows: there are three coherent signals incident, and the incident directions are 50°, 90° and 120°, respectively. In the sparse circular array, the proposed algorithm can distinguish up to three directions of arrival, and the maximum source estimation ability is also reduced compared with the non-sparse array.

Figure 4.10 shows the state of the proposed algorithm when there are four coherent sources with incident failure, and the DOA estimation has an estimation deviation of about 2°, resulting in estimation failure.

Figure 4.11 shows the 360° scanning capability of the sparse circular array in the azimuth direction. Even under the premise of sparse layout, the proposed algorithm still has the scanning ability of 360° in the azimuth direction, and the estimation ability of each direction is basically unchanged, which maintains the good characteristics of the circular array with uniformity and symmetry in the azimuth direction.

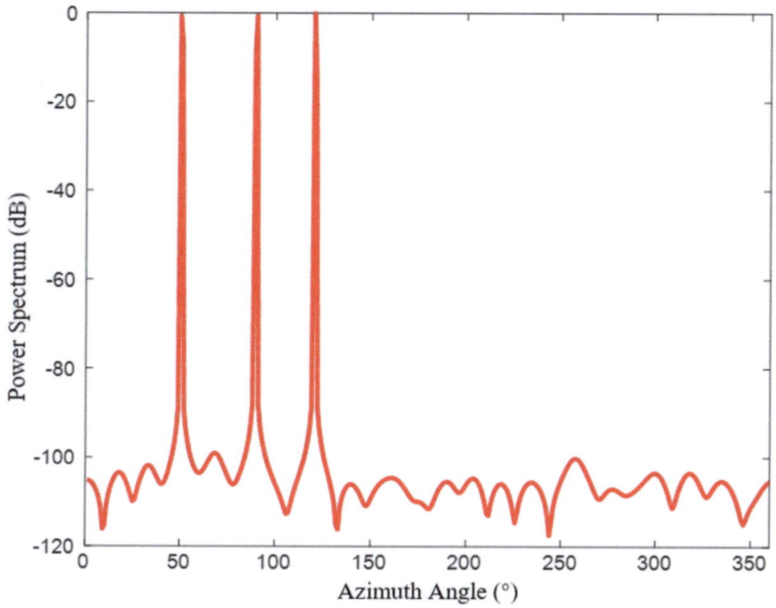

Fig. 4.9 Maximum source resolution number of sparse arrays

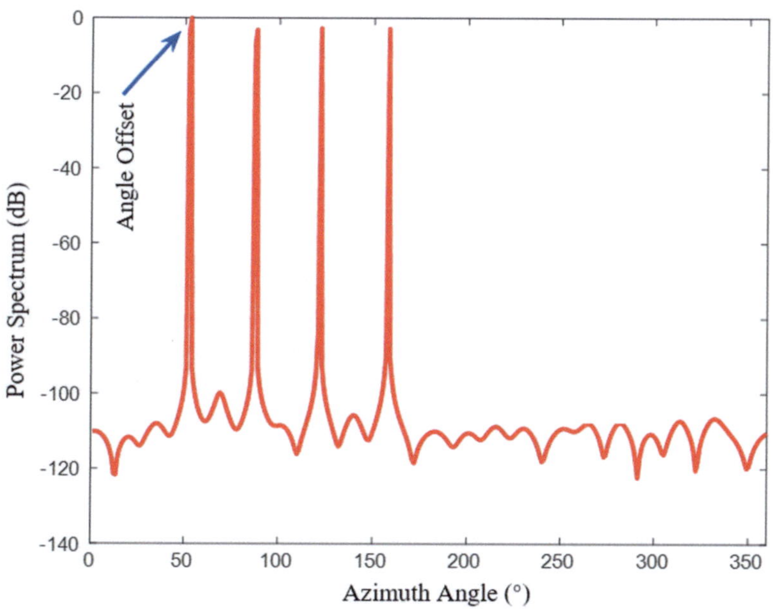

Fig. 4.10 Schematic of source resolution failure in sparse arrays

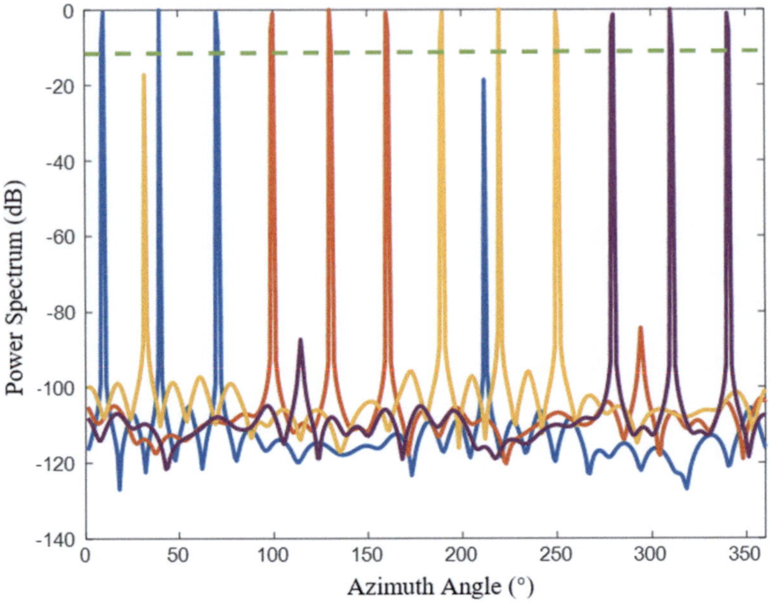

Fig. 4.11 Azimuth scanning capability of sparse circular array

4.1.2 Two-Dimensional Coherent Signal DOA Estimation Algorithm of Conformal Circular Arrays

Due to its special symmetrical planar array structure, the uniform circular array can have the same DOA estimation performance in all horizontal directions. It can provide 360° omnidirectional estimation in azimuth angles and also offer the ability to estimate pitch angles like a two-dimensional planar rectangular array for two-dimensional DOA estimation. When there is a coherent relationship among multiple received signals, operations such as decoherence need to be performed on the received different coherent signals. Although the L1-SVD algorithm can perform two-dimensional coherent signal DOA estimation of uniform circular arrays, its algorithm has high complexity, large amount of calculation, and slow calculation speed, and is not suitable for application fields such as emergency rescue that require fast calculation of DOA estimation.

Spatial smoothing algorithms are currently a mainstream, simple and effective signal decoherence algorithm. By dividing the submatrix and averaging the covariance matrix of the submatrix, the problem of rank reduction of the covariance matrix for coherent signal reception is solved. The calculation speed of this algorithm is fast, the estimation accuracy is high, and the decoherence ability is strong. However, the design of this algorithm is based on the uniform linear array, which requires that the direction vector of the antenna array signal needs to have the Vandermonde property like the uniform linear array. The received signal direction vector of a uniform circular array does not have Vandermonde properties, resulting in the inability to perform decoherence operations using spatial smoothing. Moreover, due to its simple one-dimensional spatial structure, a uniform linear array cannot simultaneously perform DOA estimates in the azimuth and pitch directions.

The mode space algorithm is a method that can virtually transform the uniform circular array in practical use into a uniform linear array. This algorithm takes the mode space as the medium, so that the signal direction vector of the transformed uniform circular array has Vandermonde properties, and the spatial smoothing algorithm can be used for decoherence. Furthermore, since the covariance matrix received by the array belongs to the complex number domain, there exist problems of large computational load and high computational complexity. In this study, the real beam constraint method is adopted to eliminate the imaginary part of the covariance matrix of the received signal, so that the two-dimensional spectral peak search process can be operated in the real number domain, which can effectively reduce the computational load and complexity of the search algorithm.

Therefore, for the two-dimensional coherent signal DOA estimation of uniform circular arrays, the spatial information of azimuth and pitch angles can be received first by using the two-dimensional planar array properties of the uniform circular array. Then, by using the mode transformation method, the direction vector of the uniform circular array can be made to have Vandermonde properties. The spatial smoothing algorithm is adopted to perform the decoherence operation on the transformed circular array. Finally, the real beam algorithm is used to simplify the process of the two-dimensional spectral peak search algorithm to find the azimuth and pitch positions of the incoming wave direction and complete the two-dimensional coherent signal DOA estimation.

4.1.2.1 Two-Dimensional Coherent Signal DOA Estimation Algorithm of Conformal Circular Arrays

(1) Mode space Algorithm

The mode space algorithm first assumes that there are D signals with the frequency of w_0, respectively at angles of (ϕ_1,θ_1), (ϕ_2,θ_2), …, (ϕ_d,θ_d) is incident into a uniform circular array with the number of elements M and the radius of the circle R. The signal received by the uniform circular array can be expressed as follows

$$
\begin{aligned}
x(l) &= \sum_{d=1}^{D} a(\phi_d, \theta_d)s_d(l) + n(l) \\
&= As(l) + n(l)
\end{aligned}
\tag{4.25}
$$

Taking the center O as the reference point, the time delay of the m-th array element relative to the reference point O is

$$\begin{aligned}\tau_m &= -\frac{1}{c}(R\cos\gamma_m \sin\theta\cos\phi + R\sin\gamma_m \sin\theta\sin\phi) \\ &= -\frac{R\sin\theta\cos(\phi-\gamma_m)}{c}\end{aligned} \tag{4.26}$$

Then the output of the m-th array element is

$$x_m = \sum_{d=1}^{D} \mathrm{e}^{-\mathrm{j}w_0\tau_{md}} s_d \quad m = 1, \ldots, M \tag{4.27}$$

According to Formula (4.27), perform the Fourier transform on the output of all array elements, and we have

$$\begin{aligned}&v_q = \sum_{m=1}^{M} x_m \mathrm{e}^{-\mathrm{j}\frac{2\pi}{M}mq} \approx M\sum_{d=1}^{D} j^{-q} J_{-q}(-\xi_d)\mathrm{e}^{-\mathrm{j}q\varphi_d} s_d \\ &q \in [-\lfloor\xi\rfloor, \lfloor\xi\rfloor] \cap Z, M > 2\lfloor\xi\rfloor\end{aligned} \tag{4.28}$$

where **Z** represents the set of integers, $\lfloor\cdot\rfloor$ is the downward rounding function, and $J_{-q}(\cdot)$ is the Bessel function of -q order.

$$\xi_d = \frac{w_0 R\sin\theta_d}{c}; \xi = \max\left[\frac{w_0 R}{c}\right] \tag{4.29}$$

$$K = \lfloor\xi\rfloor \tag{4.30}$$

Then the mode space excitation modes of this uniform circular array are -K, -$K+1$, ..., K, there are a total of $2K+1$ excitation modes, that is, this uniform circular array can be transformed into a virtual uniform linear array with $2K+1$ array elements.

Making $u_q = v_{-q}$, the above formula is rewritten in matrix form

$$\begin{bmatrix} u_{-k} \\ \vdots \\ u_k \end{bmatrix} = M \times \begin{bmatrix} j^{-K} & 0 & 0 \\ 0 & \ddots & 0 \\ 0 & 0 & j^K \end{bmatrix} \times \begin{bmatrix} J_{-K}(-\xi_1)\mathrm{e}^{-\mathrm{j}K\phi_1} & \cdots & J_{-K}(-\xi_D)\mathrm{e}^{-\mathrm{j}K\phi_D} \\ \vdots & \ddots & \vdots \\ J_K(-\xi_1)\mathrm{e}^{-\mathrm{j}K\phi_1} & \cdots & J_K(-\xi_D)\mathrm{e}^{-\mathrm{j}K\phi_D} \end{bmatrix} \times \begin{bmatrix} s_1 \\ \vdots \\ s_D \end{bmatrix} \tag{4.31}$$

that is

$$U = MJ\tilde{A}S \tag{4.32}$$

$$U = \left[u_{-K}, \ldots, u_K\right]^{\mathrm{T}} \tag{4.33}$$

$$J = diag\left\{j^{-K}, j^{-K+1}, \ldots, j^{K}\right\} \tag{4.34}$$

$$\tilde{A} = \begin{bmatrix} J_{-K}(-\xi_1)e^{-jK\phi_1} & \cdots & J_{-K}(-\xi_D)e^{-jK\phi_D} \\ \vdots & \ddots & \vdots \\ J_K(-\xi_1)e^{-jK\phi_1} & \cdots & J_K(-\xi_D)e^{-jK\phi_D} \end{bmatrix} \tag{4.35}$$

According to the spatial DFT, Formula (4.28) is rewritten in matrix form, that is

$$\begin{aligned} U &= F^{\mathrm{H}}X \\ F &= \left[w_{-K}, \ldots, w_K\right] \\ w_q &= \left[e^{-j\frac{2\pi}{M}q}, \ldots, e^{-j2\pi q}\right]^{\mathrm{H}} \end{aligned} \tag{4.36}$$

Substituting Formula (4.32), we can obtain

$$Y = TX = \tilde{A}(\phi, \theta)S \tag{4.37}$$

where $T = \frac{J^{-1} \times F}{M}$ is the mode space transformation matrix and $\tilde{A}(\phi, \theta) = \begin{bmatrix} J_{-K}(-\xi_1)e^{-jK\phi_1} & \cdots & J_{-K}(-\xi_N)e^{-jK\phi_N} \\ \vdots & \ddots & \vdots \\ J_K(-\xi_1)e^{-jK\phi_1} & \cdots & J_K(-\xi_N)e^{-jK\phi_N} \end{bmatrix}$ has Vandermonde properties.

In this way, the uniform circular array with M elements is transformed into A virtual uniform linear array with $2K + 1$ elements, Y is the output signal of the virtual uniform linear array, and $A(\phi, \theta)$ is the new array flow pattern matrix.

(2) Spatial Smoothing Algorithm

Spatial smoothing algorithm is a classical signal decoherence algorithm, which can recover the full-rank covariance matrix of coherent signals by dividing the received signal covariance matrix into submatrices and summing all submatrices to get the mean value. There are many types of spatial smoothing algorithms according to the way of subarray partition, including forward spatial smoothing, backward spatial smoothing and forward and backward spatial smoothing.

As shown in Fig. 4.12, the forward spatial smoothing algorithm will divide the virtual uniform linear array with $2K + 1$ elements converted from the uniform circular array using the mode space method into P subarrays, each of which has $L = 2K +$ 2-P elements. Take the first left subarray in Fig. 4.12 as the first subarray $x_1{}^s(t)$, then the output of the p-th subarray is $x_P{}^s(t)$

$$x_p^s(t) = [x_p x_{p+1} \ldots x_{p+L-1}] \tag{4.38}$$

The covariance matrix of the receiving array corrected by the forward spatial smoothing algorithm is the mean of the covariance matrix of each subarray after

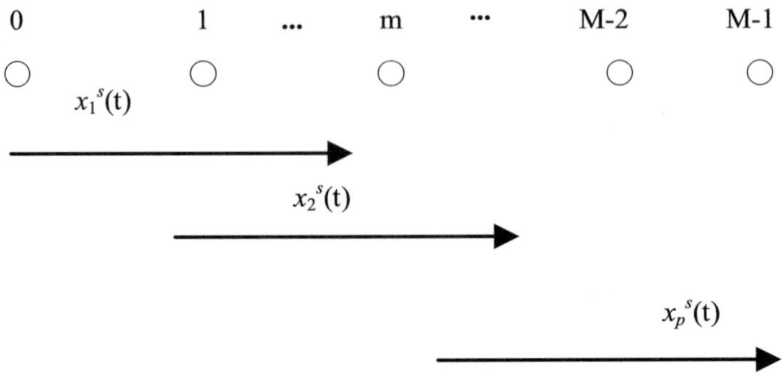

Fig. 4.12 The forward spatial smoothing algorithm

forward smoothing

$$R^s = \frac{1}{P}\sum_{p=1}^{P} R_p^s \tag{4.39}$$

where $R_p{}^s$ represents the covariance matrix of the forward spatially smoothed p-th submatrix.

Backward smoothing is to divide the subarray forward starting from the p-th element. Similarly, the covariance matrix of the receiving array corrected by the backward smoothing algorithm is as follows

$$R^b = \frac{1}{P}\sum_{p=1}^{P} R_p^b \tag{4.40}$$

where $R_b{}^s$ represents the covariance matrix of the backward spatially smoothed p-th submatrix.

According to the obtained forward spatial smoothing correction covariance matrix and backward spatial smoothing correction covariance matrix, the receiver array covariance matrix corrected by forward and backward spatial smoothing can be expressed as follows

$$R^d = \frac{R^s + R^b}{2} \tag{4.41}$$

(3) Real Beam Algorithm

The real beam algorithm is based on the mode space transformation of the uniform circular array, and the simplified calculation is done in the beam domain formed by the array receiving matrix, and the real number operation is processed. The key idea of the algorithm is that the received signal covariance matrix in the beam domain is theoretically real when the far-field narrowband uncorrelated signal is incident, and only the real part of the beam domain covariance matrix of the received signal can be retained. The algorithm constructs the signal subspace and noise subspace by matrix operations in real number domain to recover the DOA of the incident signal. The array direction vector in the uniform linear array transformed from uniform circular array has the central Hermit-symmetric property, which is convenient for covariance matrix operations in real number domain. The specific steps of the real beam algorithm are described below.

Through a central Hermitian structure of the matrix W^H left multiplication of the mode space transformation matrix, the beam space array direction vector is realized real number, and the corresponding beamforming matrix is unitary transformation, as follows

$$F_r^H = W^H T \tag{4.42}$$

The transformed beam space array direction vector is as follows

$$b(\phi, \theta) = F_r^H a(\phi, \theta) \tag{4.43}$$

Accordingly, the transformed received signal matrix is as follows

$$y'(t) = F_r^H X(t) \tag{4.44}$$

For the obtained beam space data covariance matrix, take the real part $\mathrm{Re}(R_Y)$, and calculate all the eigenvalues in the real number domain. Arrange them from large to small, and construct the noise subspace matrix G, and obtain the spatial spectrum function

$$S_b(\phi, \theta) = \frac{1}{b^T(\phi, \theta) G' G b(\phi, \theta)} \tag{4.45}$$

Based on the algorithm of mode space, spatial smoothing and real beam algorithm, a two-dimensional coherent signal DOA estimation algorithm of uniform circular array is proposed in this subsection. It can estimate two-dimensional coherent signal DOA easily and effectively, and is more suitable for the fast application field of DOA estimation.

The calculation process of the proposed algorithm is as follows.

(1) A virtual uniform linear array is obtained by changing the mode space according to the number of array elements

$$Y = TX = \tilde{A}(\phi, \theta) S \tag{4.46}$$

(2) The covariance matrix of the obtained virtual uniform linear array is spatially smoothed, and the spatially smoothed covariance matrix is obtained as follows

$$R^f = \frac{1}{P}\sum_{p=1}^{P} R_p^f \tag{4.47}$$

(3) According to the obtained covariance matrix, the received data matrix in real beam space is obtained, and its eigenvalues are arranged from large to small to obtain the noise subspace, so as to construct the spatial spectrum function

$$S_b(\phi,\theta) = \frac{1}{b^T(\phi,\theta)G'Gb(\phi,\theta)} \tag{4.48}$$

(4) Then, the azimuth angle and pitch angle information of the incident signal were obtained by two-dimensional spectral peak search in the real beam spectrum.

4.1.2.2 Two-Dimensional Coherent Signal DOA Estimation Algorithm and Simulation Results of Conformal Circular Arrays

In this subsection, the simulation test of the proposed algorithm is carried out. Under the condition that the SNR is 20 dB, the number of elements in the uniform circular array is 25, and the spacing between proximity elements is kept $\lambda/2$, Fig. 4.13 shows the maximum number of sources that can be estimated by the proposed algorithm, and Fig. 4.14 shows the three-dimensional positioning situation. The incidence direction of the signal is (100°, 20°), (20°, 30°) and (60°, 40°), including (100°, 20°), (20°, 30°) two coherent signals and (60°, 40°) one incoherent signal. The algorithm can accurately estimate the azimuth and pitch angle of the signal. And it realizes the simultaneous coherent signal and incoherent signal estimation, helps to distinguish multiple coherent targets, provides accurate target bearing information, and improves the target detection and tracking ability, which has good application value.

Figure 4.15 shows the resolution of proximity azimuth angles in the azimuth direction of the algorithm proposed in this study. The simulation conditions are as follows: two coherent signals are incident simultaneously, with the pitch angles kept the same. The incident angles are (100°, 30°) and (104°, 30°) respectively, and other conditions remain unchanged. When the power of proximity signals decreases by approximately 3dB, the azimuth proximity angle resolution of the proposed algorithm is approximately 4°.

Figure 4.16 shows the resolution of the proximity pitch angle in the pitch direction of the algorithm proposed in this study. The simulation conditions are as follows: Two coherent signals are incident simultaneously, keeping the azimuth angles the same. When two coherent signals are incident, the incident angles are (100°, 28°), (100°, 30°), and other conditions remain unchanged. When the power of proximity signals decreases by approximately 3dB, the pitch angle resolution of the proposed algorithm is approximately 2°.

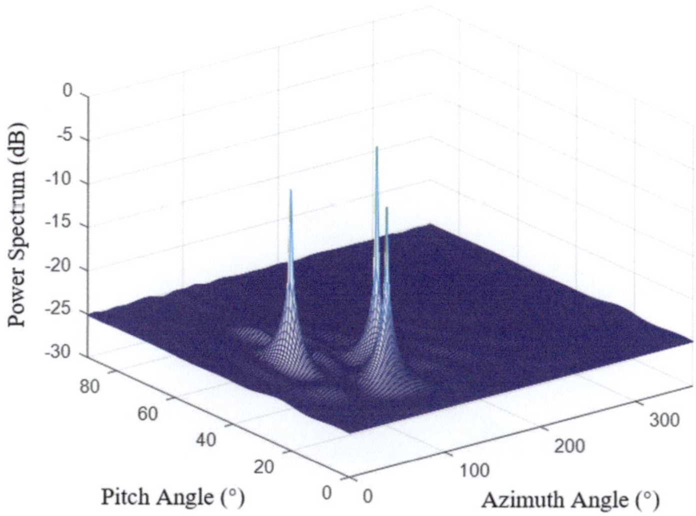

Fig. 4.13 Schematic of two-dimensional angle estimation effect

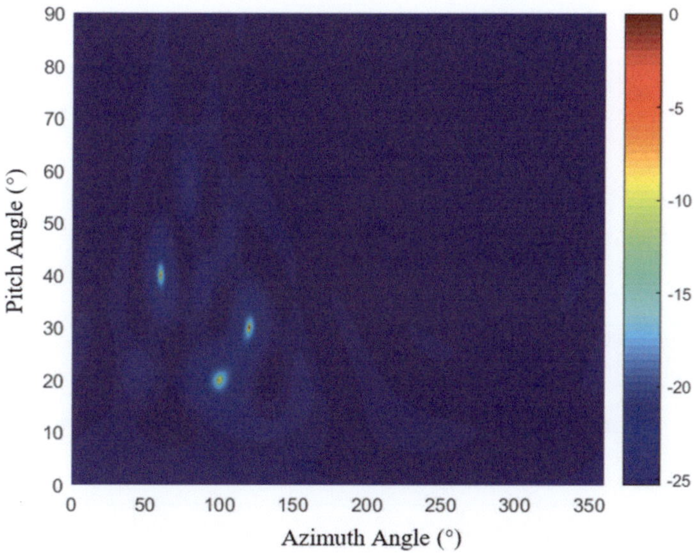

Fig. 4.14 Schematic of angle positioning

Figure 4.17 shows the pitch proximity angle estimation ability of the algorithm proposed in this study when the pitch angle changes. Keeping other conditions unchanged, when the pitch angle is close to 0°, the incident angles are (100°, 5°), (100°, 7°), and the pitch angle resolution performance remains basically unchanged.

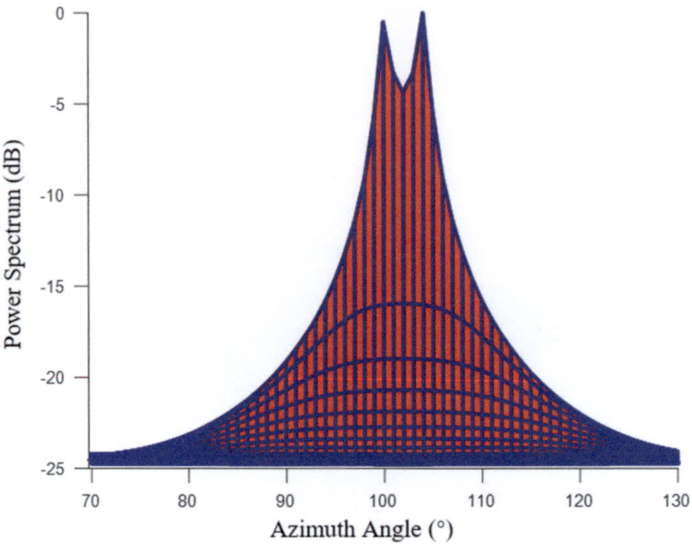

Fig. 4.15 Schematic of azimuth angle proximity angle resolution

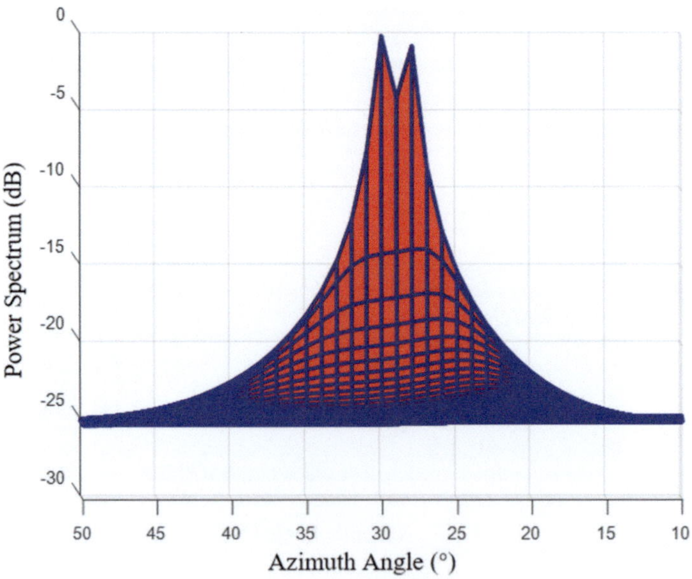

Fig. 4.16 Schematic of pitch angle proximity angle resolution

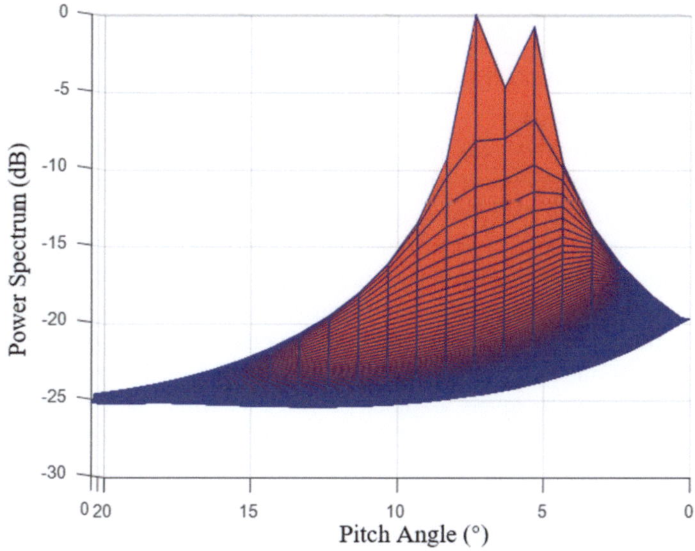

Fig. 4.17 The pitch angle proximity angle resolution when the pitch angle is close to 0°

4.2 Sparse Optimization Algorithm for Conformal Arrays

In mountainous terrains and dense forests, wireless communication usually faces various challenges. Considering the issues associated with high path loss caused by the complex terrain and diverse vegetation, severe multipath fading and large blind areas, aerial communication carriers are essentially required to improve the link connectivity and stability. In recent years, the balloon or UAV based communication has attracted increasing attention due to its relatively high maneuverability and low cost [2]. Air-ground communication offer significant benefits, such as reduced channel fading, increased coverage range and allowing cellular-mode large-capacity communication, which can be leveraged to address those challenges in mountainous and forest regions. To exploit the huge potential of air-ground communication in adverse environment, the balloon or UAV based conformal array beamforming has emerged as a highly promising technology for improving the wireless communication performance [3].

Conformal antenna arrays have the great advantage of being able to adapt to the external shapes of the aerial platforms, which not only offer good aerodynamic properties meanwhile reducing the cost and space requirements for antenna installation [4], but also help to increase the transmission distance, signal coverage range and communication capacity [5]. However, despite that conformal antenna arrays may find huge application potential in air-ground communications [6, 7], the dramatically increasing number of the array elements for high gain antennas may overload the platform, thus increasing the cost and complexity of communication systems. On the other hand, some conformal array antennas suffer from relatively high mutual coupling and sidelobe levels, which leads to unwanted energy leakage and increased interferences. To overcome these shortcomings, sparse antenna arrays [8] can be designed to optimize the radiation characteristics while reducing the weights and cost of the antenna systems.

Sparse conformal arrays are nonuniform array antennas that utilize relatively fewer radiating elements in a certain arrangement [9, 10]. For multi-dimensional structured sparse arrays, two-dimensional DOA estimation algorithms with superior accuracy are developed for source identifiability enhancement [11, 12]. Apart from DOA estimation, beamforming design is another significant task for sparse array signal processing [13]. Generally, the beamforming design of a sparse conformal array should fulfill some criteria, i.e., ensuring that the main beam gain in the desired direction remains unchanged [14], reducing the number of the array elements, and suppressing the sidelobe levels as much as possible when the element number is determined [15]. Such technical requirement imposes considerable challenges on sparse conformal array designs [16], which requires more advanced beamforming optimization algorithms. The matrix pencil algorithms [17, 18] can achieve relatively high sparse rate for linear arrays, while suffering from higher sidelobe levels and lower efficiencies. Also, the compressive sensing technique [19] is reported for sparse planar array optimization. However, these algorithms may not be very suitable for designing sparse conformal arrays due to their limited capability of dealing with nonlinear control issues. Group optimization algorithms, such as the genetic algorithm (GA) [20, 21] and particle swarm optimizer (PSO) [22, 23], are more robust in tackling nonlinear constraints and achieving optimal array arrangement for sparse conformal array designs. However, these Group optimization algorithms are computationally effortful and of slow convergence [24].

In this study, an improved snake optimization (ISO) algorithm is presented for the beamforming design of sparse conformal arrays. By combining the Sobol sequences, Cauchy mutation operator, nonlinear time-varying factor and modified flag control function, the proposed ISO algorithm can effectively acquire the best sparse array configuration as well as enhance the array sparseness and radiating performance of the conformal array antennas. Simulation verification of a sparse cylindrically conformal array is conducted. The results show that the proposed ISO algorithm achieves an array sparsity of 50% and the lowest sidelobe levels (i.e., − 14.6 dB and − 15.85 dB in the azimuth and pitch planes, respectively) among four optimization algorithms. Moreover, the convergence rate of ISO algorithm is 10% and 46% faster than the traditional snake optimization (SO) and GA, respectively. The beam steering ranges of the sparse conformal array in both azimuthal and pitch planes are also examined. Simulation results have shown the superior performance of the proposed ISO algorithm and its effectiveness in sparse conformal array beamforming designs.

4.2.1 Principles of Array Radiation Pattern Synthesis

4.2.1.1 Radiation Pattern Synthesis of Arbitrary Arrays

An array antenna is an integrated system composed of multiple independent antenna elements arranged in a specific geometric configuration to achieve targeted performance characteristics, such as directivity and beam gain. The spatial variations in the electromagnetic radiation properties of the antenna array are described by its far-field radiation pattern, the shape of which is determined collectively by factors including the number of array elements, the spacing between proximity elements, the overall geometric distribution of the elements, the amplitude and phase excitation of each element, and mutual shadowing effects among elements. By systematically and purposefully controlling these parameters, the radiation characteristics of the array can be altered and tailored to meet specific application requirements. The calculation of the array radiation pattern follows the principle of superposition: by utilizing the known radiation pattern functions of individual elements, their spatial coordinates on the carrier surface, and the expected beam direction or performance objectives, the far-field radiation pattern of the entire array can be synthesized. The computational process is outlined as follows:

Firstly, suppose there exists an antenna array composed of K elements, where the far-field radiation pattern function of the k-th element is represented as $f_k(\varphi, \theta)$. Then, the far-field radiation pattern function of the entire antenna array is given by

$$F(\varphi, \theta) = \sum_{k=1}^{K} f_k(\varphi, \theta) S_k \tag{4.49}$$

This chapter assumes that all elements in the antenna array are equal-amplitude omnidirectional elements, meaning their far-field radiation pattern functions are identical and denoted as $f(\varphi, \theta)$. Finally, under this assumption, Formula (4.49) can be simplified to

$$F(\varphi, \theta) = f(\varphi, \theta) S \tag{4.50}$$

where $f(\varphi, \theta)$ represents the far-field radiation pattern function of the individual element, and S denotes the array element position function, which depends on the spatial arrangement of the antenna elements within the array. The far-field radiation pattern of the entire array is obtained by multiplying the element's far-field radiation pattern function by the array element position function.

4.2.1.2 Radiation Pattern Synthesis of Conformal Arrays

A conformal antenna array refers to an antenna system in which individual elements are mounted on the surface of a carrier platform such that the array conforms to the shape of the surface of the carrier platform. The radiating performance of a conformal array is evaluated via its far-field radiation pattern function, which determines key beam characteristics including direction, beamwidth, and gain. By adjusting parameters such as the amplitude and phase excitation of the array elements, precise control over the beam can be achieved. Conformal arrays significantly reduce the spatial footprint of antenna systems, enhance platform aesthetics, expand beam scanning coverage, and simplify installation by minimizing mechanical complexities between the antenna and carrier. Compared to platforms equipped with non-conformal arrays, those using conformal arrays exhibit improved maneuverability due to reduced drag and structural interference. These advantages make them widely applicable in vehicles, aircraft, naval vessels, missiles, and satellites. Consequently, the design and analysis of conformal arrays have become a critical focus in antenna engineering for advanced dynamic applications.

Circular arrays are configurations where antenna elements are uniformly distributed around a circumference with a fixed radius centered at a single point. Extending this concept, a cylindrical array is formed by vertically stacking multiple identical circular arrays in space. Compared to linear arrays, circular and cylindrical arrays offer distinct advantages, including broader research potential and higher practical utility. Circular arrays inherently enable 360° azimuthal beam steering, while maintaining consistent main lobe beamwidth during scanning. Cylindrical arrays are particularly suited for aerial platforms such as multirotor/fixed-wing UAVs, rockets, or airships due to their natural structural compatibility, minimizing aerodynamic interference to preserve flight performance. Their open cylindrical geometry also expands the beam scanning coverage. In this chapter, we focus on conformal array antennas by first deriving the radiation pattern functions for conformal circular and cylindrical arrays, accompanied by analytical formulas.

As shown in Fig. 4.18, a circular array model is established with radius R, consisting of M equally spaced isotropic antenna elements uniformly distributed on the circumference. All elements share identical excitation amplitudes and omnidirectional radiation patterns. Assuming the expected main beam direction is (φ_0, θ_0), the far-field radiation pattern function toward (φ, θ) after phase compensation and beamforming can be derived from array theory and the pattern multiplication principle, expressed as

$$F(\varphi, \theta) = \sum_{m=0}^{M-1} e^{j\frac{2\pi}{\lambda}R\left[\sin\theta\cos(\varphi-\gamma_m)-\sin\theta_0\cos(\varphi_0-\gamma_m)\right]} \tag{4.51}$$

where λ is the wavelength of the transmitted signal, and γ_m denotes the angle between the position of the m-th array element and the x-axis.

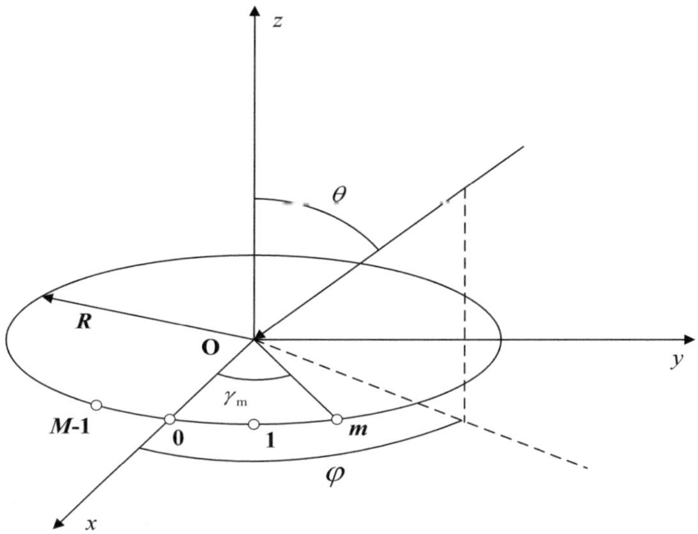

Fig. 4.18 Schematic of the structure of the uniform circular array pattern function

4.2.2 Optimization Model of the Sparse Cylindrically Conformal Array

Figure 4.19 shows the geometric model of the sparse cylindrically conformal array. Clearly, the conformal array is comprised of M ($m = 1, 2, \ldots, M$) circular arrays with the same radius of R, and each circular array consists of N ($n = 1, 2, \ldots, N$) radiating elements. For the non-sparse array, all elements on the cylindrical surface are uniformly distributed and spaced by half wavelength along the z axis and the M circles. For the sparse array, some radiating elements (i.e., the empty points) are removed through optimization and only those dark points reserved, as shown in Fig. 4.19. To simplify the simulation analysis, all radiating elements are assumed to be ideal point sources with omnidirectional radiation pattern [25] and identical excitation magnitude. It should be mentioned that the proposed algorithm in this work can be extended to accommodate more complicated sparse conformal array model, where those ideal point sources are replaced by real antenna elements and with unequal excitation power. However, those considerations are not the major points of this work, which mainly focus on advancing the optimization algorithm and validating its superiority for sparse conformal array designs.

According to the principle of in-phase field superposition for beam sharpening, taking required phase compensation for all radiating elements can readily produce a desired main beam direction toward (θ_0, ϕ_0). Correspondingly, the radiation pattern function for the sparse conformal array as shown in Fig. 4.19 can be expressed as:

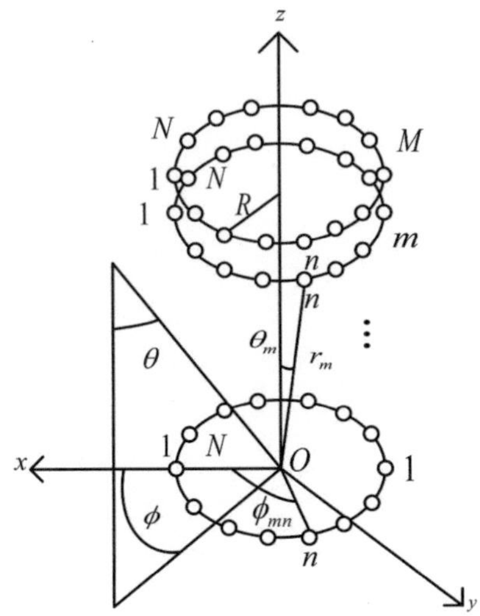

Fig. 4.19 Geometric model of the sparse cylindrically conformal array

$$
\begin{aligned}
& F(\theta, \phi; \theta_0, \phi_0) \\
& = \sum_{m=0}^{M-1} \sum_{n=0}^{N-1} f_{\mathrm{mn}} \cdot \exp\Big\{ \mathrm{j}\frac{2\pi}{\lambda} r_m [\sin\theta_m \sin\theta \cos(\phi - \phi_{mn}) \\
& \quad - \sin\theta_m \sin\theta_0 \cos(\phi_0 - \phi_{mn}) + \cos\theta_m (\cos\theta - \cos\theta_0)] \Big\}
\end{aligned} \tag{4.52}
$$

where λ is the wavelength of the electromagnetic wave, (θ, ϕ) is the direction of an arbitrary field point, (θ_0, ϕ_0) is the desired maximum radiating direction, $(\theta_{\mathrm{m}}, \phi_{\mathrm{mn}})$ is the direction of the nth source element in the m-th circular array, r_{m} is distance between the coordinate origin O and an arbitrary source element in the m-th circular array. In addition, the factor f_{mn} describes the sparsity of the conformal array, where $f_{\mathrm{mn}} = 1$ or 0 indicates that there is an or no element at the corresponding position of the conformal array.

According to the definition of the maximum sidelobe level (MSLL) of the radiation pattern, the calculation formula of MSLL can be expressed as

$$
MSLL = \max_{\phi \in S}\{F_{\mathrm{norm}}(\phi)\} = \left| \frac{\max_{\phi \in S}\{F(\phi)\}}{\max\{F(\phi)\}} \right| \tag{4.53}
$$

where $F_{\mathrm{norm}}(\phi)$ represents the normalized radiation pattern function and S represents the angle set corresponding to all sidelobes of the radiation pattern.

According to the design criteria of a sparse conformal array, the optimization model can be defined as $\min_{f_{mn}}(MSLL)$, where the constraint conditions are expressed as follows

$$E(u_0) = 1$$
$$p = q \tag{4.54}$$

The physical meaning of optimization model is to find the optimum array configuration or the f_{mn} values for minimizing the sidelobe level in the radiation pattern, under the constraints of a certain sparsity rate and normalized main-lobe level. In Formula (4.54), $E(u_0)$ is the normalized electric field strength along the desired main-beam direction, p is the variable sparsity rate and q is the required sparsity value.

4.2.3 Improved Snake Optimization Algorithm and Simulation Results

The proposed ISO algorithm is an improved version of the traditional SO. As a group optimization algorithm, the traditional SO algorithm [26] is inspired by the behaviors of the snakes and consists of two stages, i.e., global exploration and local exploitation. In the global exploration stage, since there is no food nearby, the snakes search for food by moving closer or farther away from certain individuals. Once enough food is obtained, the SO algorithm enters its local exploitation stage that includes several transition stages to increase the optimization efficiency. Despite being an effective algorithm for solving complex problems, the SO algorithm has some limitations, such as the possibility of sliding into local optima due to its randomly generated population and long iteration processes that consume much computing time [27]. Additionally, the algorithm may obtain approximate and imprecise solutions that are not ideal. To overcome these limitations, this work proposes the ISO algorithm with three major improvements to the traditional SO.

4.2.3.1 Sobol Sequence Initialization

To improve the initialization process of SO, the conventional algorithm of generating a uniformly distributed random population (i.e., the initial array layout set) is replaced by using the Sobol sequence. The Sobol sequence is a deterministic quasi-random number sequence that provides low-variance and uniform points for the population initialization process [28]. Unlike pseudo-random number sequences, Sobol sequences use radical inversions with base 2 in each dimension of the population, and each dimension has its own unique matrix, resulting in non-repeating and uniform points. This approach can achieve better array layout distribution and higher travers ability, which ultimately leads to better optimization results.

For brevity, the i-th initial individual is described as

$$X_i = X_{\min} + S_i \times (X_{\max} - X_{\min}) \tag{4.55}$$

where $X_{\max}$ is the upper bound, $X_{\min}$ is the lower bound and S_i is the ith random number generated by a Sobol sequence that is distributed in the interval of (0, 1). From the generated initial array layout set, the current optimal array layout X_{current} is selected by judging and evaluating the calculated fitness values.

4.2.3.2 Cauchy Mutation Strategy

To enhance the exploration capability of the SO algorithm, we propose the Cauchy mutation strategy in combination with Cauchy distribution [29]. Provided an individual (i.e., the array layout) was trapped in a local optimal solution for many times that exceed the limit, a Cauchy perturbation would be applied to the current optimal individual. The Cauchy operator produces a big step, allowing the algorithm to jump out of the local optima. Additionally, we introduce smaller phases to accelerate the search for the best solution. For brevity, the Cauchy mutation strategy is described as

$$f_{\text{cauchy}}(x; x_0, \gamma) = \frac{1}{\pi} \times \left[\frac{\gamma}{(x - x_0)^2 + \gamma^2} \right] \tag{4.56}$$

where $f_{\text{Cauchy}}(x; x_0, \gamma)$ is the probability density function of Cauchy distribution, x *is* a random variable subject to the Cauchy distribution, x_0 is the location parameter that defines the peak position, γ is the scale parameter corresponding to half the width of the probability density function at half the maximum height. For the case of x_0 equal to 0 and γ equal to 1, the Cauchy distribution can be expressed as

$$f_{cauchy}(x; 0, 1) = \frac{1}{\pi \times (x^2 + 1)} \tag{4.57}$$

Utilizing the Cauchy mutation strategy yields

$$X_{\text{new}} = X_{\text{current}} \times [1 + cauchy(0, 1)] \tag{4.58}$$

where X_{new} is the new array layout obtained after Cauchy perturbation, and *Cauchy*(0,1) is the Cauchy operator that creates a random variable subject to Cauchy distribution as Formula (4.57).

Once execution of the Cauchy mutation strategy is completed, the ISO algorithm enters the exploration or exploitation stage.

4.2.3.3 Nonlinear Time-Varying Factor and Dynamic Adaptive Weighting

During the exploration or exploitation stage, the algorithm updates the array layout. The traditional SO algorithm randomly updates the individual population, which inevitably lead to slow convergence and out of control. To tackle this issue, a new population control update function is developed in this work. Inspired by the whale algorithm, this new function utilizes a nonlinear time-varying function, which enables better adjustment of global exploration and local development. For brevity, the proposed population control update function is described as

$$U_{flag} = 2 \times a \times r - a \tag{4.59}$$

and

$$\begin{cases} a = 1 + \cos\left(\pi \times \frac{t}{T}\right), & t \leq 0.5T \\ a = \cos\left(\pi \times \frac{t-0.5T}{T}\right), & t > 0.5T \end{cases} \tag{4.60}$$

where r is a uniformly random function, T is the total iteration number, and t is the current number of iterations.

Provided U_{flag} is less than 1, the worst male and female individuals are selected and replaced. However, it is essentially time-consuming to perform random updating of the worst individuals. To tackle this issue, dynamic adaptive weighting is proposed and developed to accelerate updating process and improve the algorithm convergence speed. For brevity, the dynamic adaptive weight is described as

$$w = \tanh[2 \times (1 - t/T)] \tag{4.61}$$

and the worst individual updating function is described as

$$X_{\text{worst}} = X_{\min} + w \times r \times (X_{\max} - X_{\min}) \tag{4.62}$$

where w is the dynamic adaptive weight, and X_{worst} represents for the worst male and female individuals. Once completion of the updating processing, the fitness values are recalculated. The best individuals from the male and female populations are compared as per their fitness values, and the individual with the highest fitness value is chosen as the current optimal array layout X_{current}. Once reaching the maximum iteration time, the finally updated array layout X_{current} is the output of the ISO algorithm. Figure 4.20 shows the specific flowchart of the proposed ISO algorithm. In addition, the pseudo code of the proposed ISO algorithm is shown in Algorithm 1.

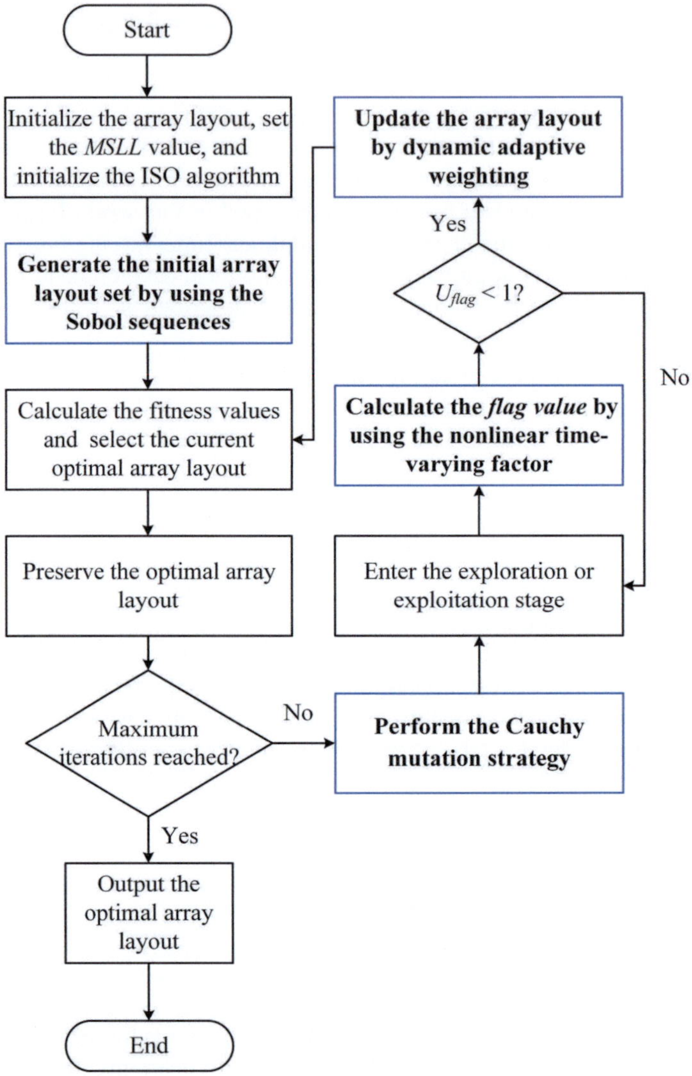

Fig. 4.20 The flowchart of the proposed ISO algorithm

Algorithm 1 ISO Algorithm

Initialize the array layout set with sobol sequences and define relevant parameters
Calculate the fitness *MSLL* of each array layout and obtain the current optimal array layout X_{current}
Divide the population N into two equal numbers of N_{M} and N_{F}, and find the optimal male and female individuals
while $t < T$ **do**
if limit > LIMIT **then**
Carry out Cauchy mutation to generate X_{new}
end if
Calculate the values of K and Q
if $Q < 0.25$ **then**
enter the exploration stage
else if $K > 0.6$ then
enter the exploitation stage
else
if *rand* > 0.6 then fight
else
enters mating mode
Calculate U_{flag}
if $U_{flag} < 1$ then
Select the worst male and female individuals, and
replace them by dynamic adaptive weighting
end if
end if
end if
Calculate the fitness *MSLL* of the new array layout set
Update the current optimal array layout if there is a better solution
$t = t + 1$
return the optimal array layout

4.2.3.4 Simulation Results and Analysis

Considering a cylindrical conformal array with $M = 20$, $N = 25$, $\lambda = 1$ m and $R = 2\lambda$, the desired array sparsity is 50%. To compare the algorithm performance, beamforming simulation for the conformal array is carried out by using four different optimization algorithms. For the proposed ISO algorithm, the size of the initial array layout set is set to 50, and the maximum number of iterations is 200. For the SO algorithm, $c1 = 0.5$, $c2 = 0.05$ and $c3 = 2$; For the PSO algorithm, $c_1 = c_2 = 2$, and the maximum and minimum weights are 0.8 and 0.4, respectively; For the GA algorithm, the crossover rate is 0.8, the mutation rate is 0.2 and the selection operation is based on roulette.

Figures 4.21 and 4.22 show the simulated radiation patterns in the azimuthal and pitch planes for the non-sparse and sparse conformal arrays. Here, the required main beam direction (θ_0, ϕ_0) is set to $(90°, 81°)$, and sparse array beamforming is conducted by using the proposed ISO algorithm. Clearly, a highly directional radiating beam is

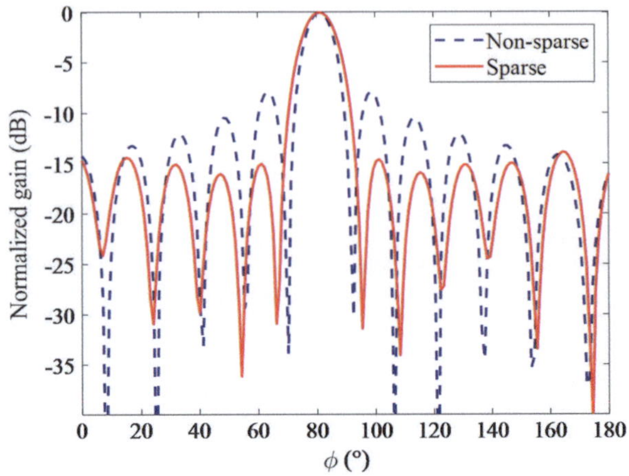

Fig. 4.21 Simulated radiation patterns in the azimuthal plane for the non-sparse and sparse conformal arrays

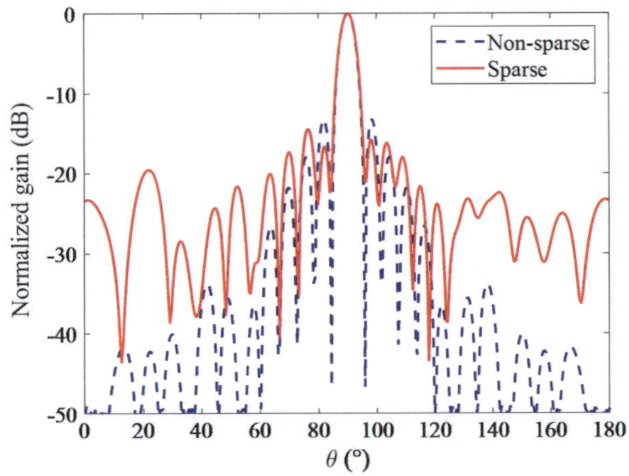

Fig. 4.22 Simulated radiation patterns in the pitch plane for the non-sparse and sparse conformal arrays

produced with the half-power beamwidths (HPBWs) of 12.6° and 5° in two principal planes, respectively. By comparing the radiation patterns of the non-sparse and sparse conformal arrays, it is found that there is no remarkable variation in the beamwidth. The beamwidth after sparseness is slightly increased by approximately 2.6° in the azimuthal plane, while unchanged for the pitch pattern. The maximum sidelobe levels in the azimuth and pitch planes are effectively reduced by 7.5 dB and 3.52 dB for a sparsity of 50%, respectively. Such improvement is believed to benefit from reduced interferences between the array elements toward and against the main beam direction. As clearly shown in Fig. 4.23, the achieved array arrangement has indicated that the number of the radiating elements are significantly reduced at those positions against the main beam direction (90°, 81°).

Figures 4.24 and 4.25 show the simulated radiation patterns for the sparse conformal array by using different beamforming optimization algorithms. Clearly,

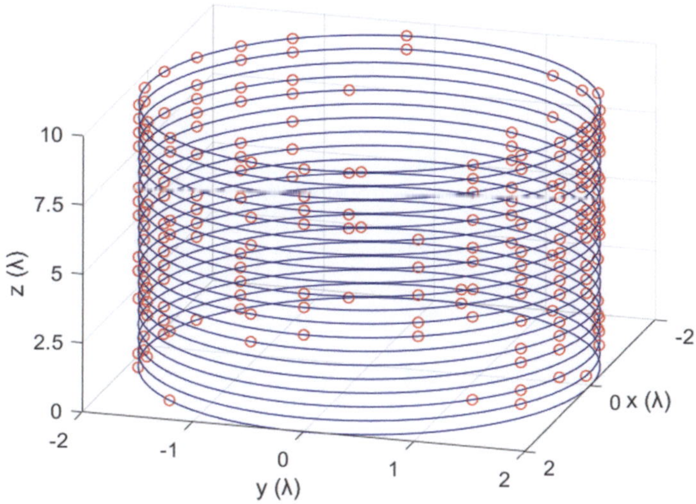

Fig. 4.23 Simulated array element arrangement for the sparse cylindrically conformal array by using ISO

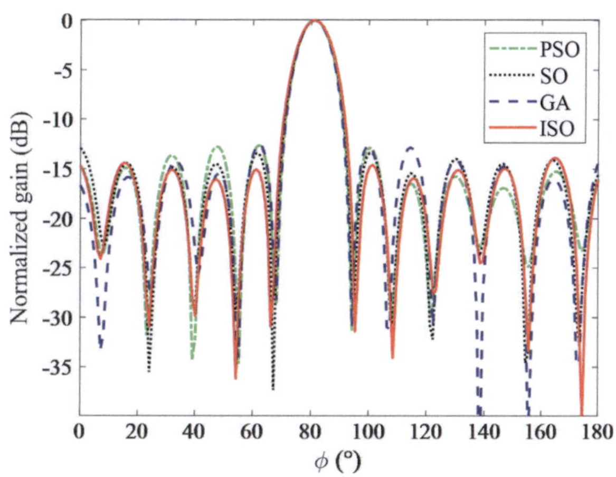

Fig. 4.24 Simulated radiation patterns in the azimuthal plane for the sparse conformal array by using different beamforming optimization algorithms

these radiation patterns exhibit almost identical beamwidths. However, the radiation patterns optimized using ISO exhibit the lowest maximum sidelobe levels, which are − 14.6 and − 15.85 dB in the azimuthal and pitch planes. As compared with SO, PSO and GA, the maximum sidelobe levels optimized using ISO are lower by 1.23 dB, 1.99 dB and 2.08 dB in the azimuthal plane, and 1.77 dB, 2.09 dB and 1.49 dB in the pitch plane, respectively.

Figure 4.26 illustrates the iteration convergence rates for beamforming optimization of the sparse conformal array by using different algorithms. Simulation results have demonstrated that the proposed ISO algorithm not only achieves the lowest

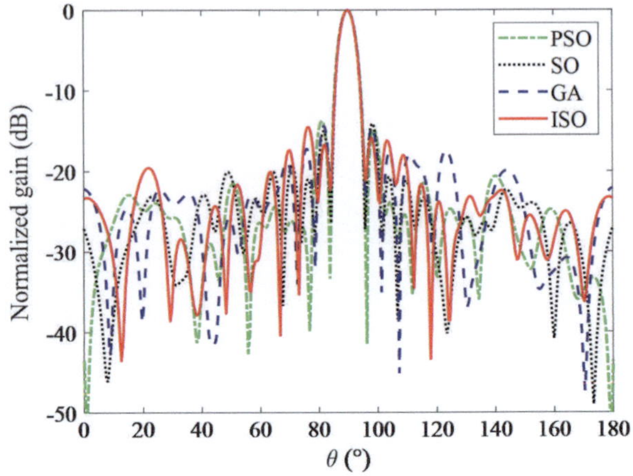

Fig. 4.25 Simulated radiation patterns in the pitch plane for the sparse conformal array by using different beamforming optimization algorithms

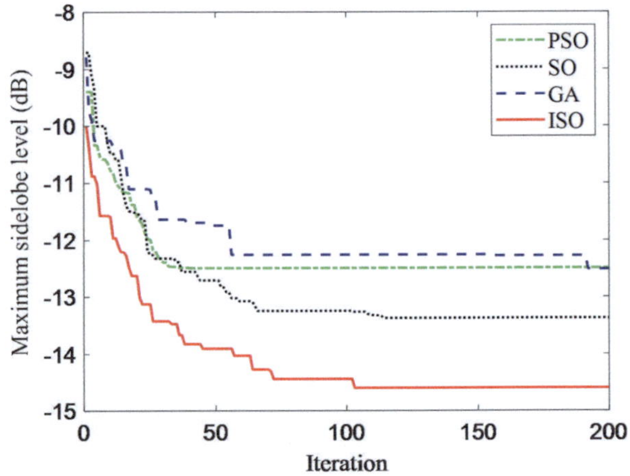

Fig. 4.26 Simulated iteration convergence rates for beamforming design of the sparse conformal array by using different optimization algorithms

maximum sidelobe level, but also converges faster than the SO and GA algorithms. More specifically, the convergence rate for ISO is 10% and 46.53% faster than that for SO and GA, respectively, but a bit slower than that for PSO. However, considering the comprehensive performance in terms of sidelobe reduction ability and iteration convergence rate, it is concluded that the proposed ISO algorithm exhibits superior performance over other optimization algorithms.

Figures 4.27 and 4.28 illustrate the radiation patterns of the sparse conformal array for beam steering in the azimuthal and pitch planes, respectively. It should be mentioned that the array element arrangement as shown in Fig. 4.23 is kept unchanged during beam steering. Clearly, the main-lobe shapes and peak gains are almost the same for different beam steering angles within 54° and 108° in the azimuthal plane. However, the MSLL gradually deteriorates with increasing the steering angle, achieving − 10 dB at two edge steering angles. For beam steering in the pitch plane,

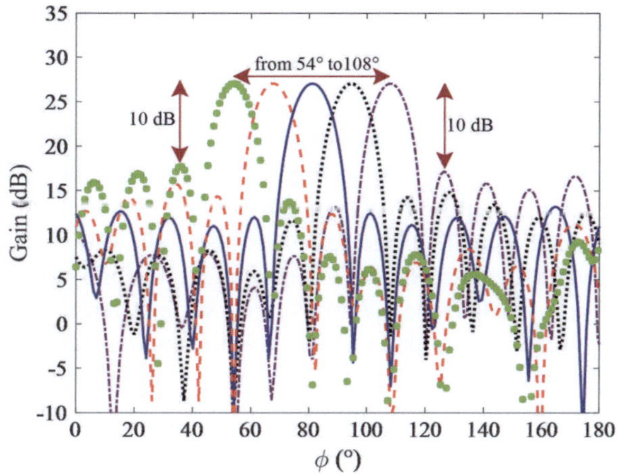

Fig. 4.27 Simulated radiation patterns of the sparse conformal array for beam steering in the azimuthal plane

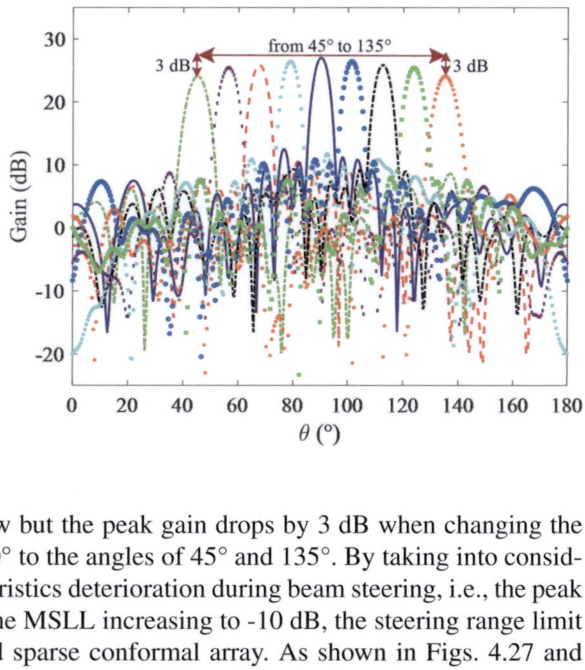

Fig. 4.28 Simulated radiation patterns of the sparse conformal array for beam steering in the pitch plane

the MSLL is always very low but the peak gain drops by 3 dB when changing the main beam direction from 90° to the angles of 45° and 135°. By taking into consideration the radiation characteristics deterioration during beam steering, i.e., the peak gain decreasing by 3 dB or the MSLL increasing to -10 dB, the steering range limit is evaluated for the designed sparse conformal array. As shown in Figs. 4.27 and 4.28, the beam steering ranges are around 54° and 90° in the azimuthal and pitch planes, respectively. It should be mentioned that the steering range limit can further be enlarged by incorporating more freedom of design in future designs, e.g., involving excitation amplitude weighting for the array elements. Finally, the aperture efficiency is calculated from the simulated gain value at the scanned angle and the overall area of the cylindrical surface covering the sparse conformal array, which is estimated to be around 33.4% at the center of the beam scanning range, about 8.7% lower than that of the corresponding non-sparse conformal array design.

4.3 Interference Suppression Beamforming Algorithm for Conformal Sparse Array

IRAB is primarily employed for anti- interference communications. It utilizes array antenna to dynamically control the beam steering over the entire receiving spatial domain, adapting to changes in the expected and interference signal directions. By forming spatially directed beams, the main lobe of the array antenna is consistently aligned with the expected signal direction, while nulls are steered toward interference directions. This mechanism suppresses interference signals, enhances the utilization of user signals, and improves the reception performance of the antenna system. Research on IRAB focuses on adaptively adjusting the excitation coefficients of individual array elements to dynamically reshape the beam pattern according to real-time variations in expected and interference directions, thereby ensuring high gain in the signal direction and deep suppression in interference directions. This ultimately enhances communication metrics such as the SINR. Current IRAB algorithms, such as the DL algorithm, face challenges including the difficulty in selecting optimal diagonal loading factors and incompatibility with conformal sparse arrays. To address these limitations, this section proposes a joint algorithm that integrates updating diagonal loading (UDL) algorithm with a combined particle swarm optimization (CPSO) algorithm. This algorithm is specifically tailored for conformal sparse arrays, aiming to overcome the shortcomings of conventional IRAB algorithms and achieve robust interference suppression performance.

This section first investigates the principle of deriving the optimal weight vector in beamforming. Guided by the maximum signal to interference plus noise ratio (MSINR) criterion, it establishes the solution method of solving optimal weight vector for IRAB. The robustness of IRAB algorithms is studied to reduce their dependency on covariance matrix of ideal received signals. By analyzing the eigenvalues of the covariance matrix, the range of the diagonal loading factor is divided. Then, an UDL algorithm for IRAB is proposed. Futhermore, a CPSO algorithm is developed, integrating three strategies: chaotic sequences, velocity weight updates, and cross and mutation. Specifically, the cross and mutation strategy solves the problem that the particle swarm algorithm is easy to fall into the local optimal solution and cannot jump out, while chaotic sequences and velocity weight updates strategies improve convergence speed and optimization accuracy. Building on these, a joint interference suppression beamforming algorithm that integrates UDL and CPSO algorithms for conformal sparse array is proposed. Following the maximum interference suppression level criterion, an innovative adaptive beamforming framework for sparse circular arrays is designed. This approach achieves effective interference suppression while reducing the number of array elements and lowering system costs.

4.3.1 Principles of Interference Suppression Beamforming

Adaptive beamforming focuses on determining the weight coefficient vector matrix of an entire array, thereby generating the expected array pattern function based on these weight coefficients. To distinguish effective expected signals from co-channel interference and noise signals in received signals, adaptive beamforming algorithms typically rely on prior knowledge such as known SNR, interference signal strength, array steering vectors and DOA of expected and interfering signals. According to the predicted information, the algorithm solves the beamforming problem to obtain the excitation coefficients of each antenna element, which is the optimal weight vector of the array.

4.3.1.1 The Optimal Weight Vector of Array

Suppose the number of elements of a certain array is M, then define w as the set of weight vectors, expressed as

$$w = [w_1, \ldots, w_M]^T \tag{4.63}$$

The received signals of each element are synthesized through the set of weight vectors to generate the array output pattern function, then the weighted output of the array at a certain time t is as follows

$$y(t) = w^H x(t) = \sum_{m=1}^{M} w_m x_m(t) \tag{4.64}$$

The output response and spatial beam pattern of an array consequently varies with different weight vectors. Assuming all array elements are isotropic, the signal amplitude of each element remains identical and can be normalized to unity 1. In this scenario, the weight vector set exclusively controls the phase differences between array elements without altering signal amplitudes.

Consider a far-field scenario where an expected signal $e(t)$ and I interference signals are transmitted. Assuming these signals incident to the array from angles θ_e (expected) and θ_i (interference, $i = 1, 2, \ldots, I$), the received signal vector $x(t)$ at the array can be modeled as

$$x(t) = As(t) + n(t) = a(\theta_e)e(t) + \sum_{i=1}^{I} a_i(\theta_i)i_i(t) + n(t) \tag{4.65}$$

where $a(\theta_k) = [a_1(\theta_k), \ldots, a_m(\theta_k)]^T$ represents the signal direction vector with the incident direction of θ_k.

According to the requirements of IRAB design, in order to enable the array to receive the expected direction signal without distortion and suppress the power of the interfering direction signal, it is necessary to adjust through the weight vector function to keep the power of the expected direction signal unchanged and eliminate the power of the interfering direction signal. Therefore, the constraint conditions of the weight vector algorithm are

$$\begin{aligned} w^H a(\theta_e) &= 1, \\ w^H a(\theta_i) &= 0 \end{aligned} \tag{4.66}$$

In addition, beamforming also needs to consider the influence caused by the superposition of environmental noise. Suppressing interference and noise needs to be carried out simultaneously. Under the condition of keeping the power in the expected signal direction unchanged, minimizing the total output power value of the array can achieve the purpose of suppressing both interference and noise simultaneously. The covariance matrix R of the array output can be expressed as three parts: signal, interference and noise

$$\begin{aligned} R &= E\left[x(t)x(t)^H\right] \\ &= E\left[s_0(t)s_0^H(t)\right] + \sum_{i=1}^{J}\left[s_i(t)s_i^H(t)\right] + E\left[n(t)n^H(t)\right] \\ &= R_s + R_i + R_n \end{aligned} \tag{4.67}$$

where R_s is the covariance matrix of the expected signal, R_i is the covariance matrix of the interference signal, and R_n is the covariance matrix of the noise signal.

The total output power P of the array is expressed as

$$\begin{aligned} P(y) = E\left[|y(t)|^2\right] &= w^H E\left[x(t)x^H(t)\right]w \\ &= w^H R_s w + w^H R_i w + w^H R_n w \end{aligned} \tag{4.68}$$

The power P_s of the expected signal, the power P_i of the interference signal, and the power P_n of the noise signal output by the array can be calculated respectively using the weight vector and the corresponding covariance matrix, and are expressed as

$$P_s = w^H R_s w \tag{4.69}$$

$$P_i = w^H R_i w \tag{4.70}$$

$$P_n = w^H R_n w \tag{4.71}$$

The output SINR of the array is expressed as the power of the expected signal divided by the sum of the power of the interference signal and the noise signal, denoted as

$$
\begin{aligned}
SINR &= \frac{P_s}{P_i + P_n} = \frac{w^H R_s w}{w^H R_i w + w^H R_n w} \\
&= \frac{w^H R_s w}{w^H (R_i + R_n) w}
\end{aligned}
\tag{4.72}
$$

After beamforming, the array output based on the principle of MSINR can be achieved by designing and selecting the appropriate optimal weight vector.

4.3.1.2 MSINR Criterion

The MSINR of the array output was used as the optimization function of the weight vector

$$
\max_{W} SINR = \max_{W} \frac{W^{\mathrm{H}} R_S W}{W^{\mathrm{H}} R_{i+n} W} \tag{4.73}
$$

where R_{i+n} is the covariance matrix of interference and noise signals.

Ensuring that the expected signal direction gain remains unchanged and under the criterion of the MSINR of the array output, the solution limit of the array optimal weight vector becomes

$$
\begin{aligned}
&\max_{W} SINR = \max_{W} \frac{W^{\mathrm{H}} R_S W}{W^{\mathrm{H}} R_{i+n} W} \\
&s.t.\ w^H a(\theta_e) = 1
\end{aligned}
\tag{4.74}
$$

The optimal weight vector can be expressed as

$$
w_o = \frac{R_{i+\mathrm{n}}^{-1} a(\theta_e)}{a^{\mathrm{H}}(\theta_e) R_{i+\mathrm{n}}^{-1} a(\theta_e)} \tag{4.75}
$$

The subsequent IRAB algorithm design and simulation in this study are all carried out based on the MSINR criterion.

4.3.2 *Interference Suppression Beamforming Algorithm for Sparse Circular Arrays*

4.3.2.1 Design and Implementation of CPSO Algorithm

The PSO algorithm draws inspiration from observations of collective movement patterns in biological groups such as bird flocks or fish schools. The algorithm mimics the information-sharing behavior among individuals in such groups to search for optimal solutions to problems, such as locating the optimal solution or minimizing errors. In PSO, candidate solutions for the problem's optimal solution are termed particles. Each particle has its own position, movement direction, and velocity. Their fitness in the solution space is determined by evaluating a fitness function, and they navigate through the solution space accordingly. Particles update their velocity and position by learning from their own best-known position and the group's best-known position. The algorithm continuously updates the particles' positions and velocities, guiding them toward the target optimal solution. PSO is characterized by its simplicity of implementation, strong parallel computing capabilities, insensitivity to the initial solution distribution, and ease of parameter tuning. However, it also suffers from limitations such as difficulty in escaping local optima and challenges in handling constrained optimization problems. The main steps of the PSO algorithm are as follows:

(1) Initialization: Generate the positions and velocities of all particles in the population randomly

Suppose the initial population number is N and the dimension of the optimization variable is D, then the initial position of the i-th particle can be expressed as

$$X_i = (x_{i1}, x_{i2}, \ldots, x_{iD}), i = 1, 2, \ldots, N \tag{4.76}$$

The initial flight velocity of the i-th particle can be expressed as

$$V_i = (v_{i1}, v_{i2}, \ldots, v_{iD}), i = 1, 2, \ldots, N \tag{4.77}$$

(2) Evaluate fitness Calculate the current fitness of each particle based on the objective function of the problem.

Record the optimal fitness position achieved by the i-th particle in the current iteration number as the current individual extreme value, denoted as

$$p_{i\text{best}} = (p_{i1}, p_{i2}, \ldots, p_{iD}), i = 1, 2, \ldots, N \tag{4.78}$$

The optimal fitness position achieved by the entire particle population under the current number of iterations is recorded as the current global extreme value, denoted as

$$g_{\text{best}} = (g_1, g_2, \ldots, g_D) \tag{4.79}$$

(3) Update positions and velocities

Particles update their velocity and position based on the current individual extremum and the current global extremum

$$\begin{aligned} &v_{ij}(t+1) = v_{ij}(t) + c_1 r_1(t)\left[p_{ij}(t) - x_{ij}(t)\right] + c_2 r_2(t)\left[g_{best}(t) - x_{ij}(t)\right] \\ &x_{ij}(t+1) = x_{ij}(t) + v_{ij}(t+1) \\ &v_{ij} \in [-v_{\max}, v_{\max}] \end{aligned} \tag{4.80}$$

where c_1 and c_2 are the particle learning parameters; r_1 and r_2 are random numbers uniformly distributed within (0,1), which are used to increase the randomness of particle flight to simulate the actual flight environment. $v_{ij} \in [-v_{\max}, v_{\max}]$ represents the particle velocity of the i-th particle in the j-th dimension. And $v_{\max}$, which is a constant and is set by the user to limit the velocity range of the particle, represents the maximum velocity of the particle.

This section combines algorithms such as cross-mutation strategy, chaotic function and nonlinear weight to conduct combinatorial optimization of the PSO algorithm. A new CPSO algorithm is proposed to help the algorithm get out of the local optimal predicament in time and optimize the initial solution distribution of the algorithm, thereby improving the computing speed.

(1) Initialization of iterative chaotic map with infinite collapse (ICMIC)

The ICMIC is used instead of the traditional algorithm of initially generating uniformly distributed random population. Through chaotic mapping, a large number of non-repeated points are generated above and below the value range, and this algorithm can achieve better boundary traversal search ability. Its mathematical expression is as follows

$$x_{k+1} = \sin(a/x_k) \tag{4.81}$$

where a is a constant distributed over the interval $(0, +\infty)$.

(2) Update strategy of velocity weight

In order to improve the convergence speed of the PSO algorithm, we propose an update strategy of velocity weights.

$$v_{ij}(t+1) = w \times v_{ij}(t) + c_1 \times r_1(t)\left[p_{ij}(t) - x_{ij}(t)\right] + c_2 \times r_2(t)\left[p_{gj}(t) - x_{ij}(t)\right] \tag{4.82}$$

$$\begin{aligned} w &= w_{\max} - \frac{(w_{\max} - w_{\min}) \times t}{T} \\ c_1 &= c_{\max} - \frac{(c_{\max} - c_{\min}) \times t}{T} \\ c_2 &= c_{\max} - \frac{(c_{\max} - c_{\min}) \times t}{T} \end{aligned} \tag{4.83}$$

where w is called the inertia weight, $w_{\max}$ and $w_{\min}$ are the value ranges of the inertia weight, $c_{\max}$ and $c_{\min}$ are the value ranges of the acceleration weight, t is the current number of iterations, and T is the maximum number of iterations. The velocity of the particle swarm will gradually decrease with the increase of the number of iterations. At the beginning of the algorithm, it conducts global exploration at a relatively high speed, and in the later stage, it conducts local search at a relatively low speed to accelerate the calculation speed and improve the optimization ability.

(3) Cross-mutation strategy

The transmission algorithm, through the crossover and mutation operations of genes, has the ability to jump out of the local optimal solution, strong global search ability and parallel information processing ability. However, the crossover and mutation operations lead to a large amount of calculation and a slow convergence speed of the GA. Considering the characteristics of particles in the PSO algorithm, such as memorability, fast convergence speed, but difficulty in escaping the local optimal solution, a particle swarm individual cross-mutation strategy is designed. This strategy assists the algorithm in jumping out of the local optimal solution, increases the flexibility of particle changes, overcomes the problem of insufficient search accuracy of the PSO algorithm, and realizes the complementary advantages among different algorithms.

After the process of updating the individual extremum and the total extremum of the particle swarm, the cross-mutation strategy is introduced. Define the genetic population *GAPOP*, whose population number is consistent with the population number N of the particle swarm. Define the population crossover number P_c and the population variation number $P_m = N - P_c$.

Generate the cross-population GA_c,

$$GA_c = 0.5 \times GA_{pc} + 0.5 \times g_{best} \tag{4.84}$$

where GA_{pc} is the crossover population selected from the genetic population.

Generate the mutant population GA_m,

$$\begin{aligned} GA_m &= GA_{pm} + GA_{rand} \\ GA_{rand} &= rand([p_{\min}, p_{\max}], P_m, D) \end{aligned} \tag{4.85}$$

where GA_{pm} is the mutant population selected from the genetic population, $p_{\max}$ is the maximum value within the particle variation range, $p_{\min}$ is the minimum value within the particle variation range, and D is the population dimension.

Finally, the population is updated to

$$POP = [GA_c, GA_m] \tag{4.86}$$

4.3.2.2 Beamforming Algorithm Based on UDL Algorithm

DL algorithm is a classic robust beamforming algorithm, which is usually used in cases where the spatial separation interval between the interference and the expected signal is far and the strength of the expected signal is weaker than that of the interference. The proposal of the DL algorithm is due to the fact that noise disturbance leads to a slower convergence speed of the sample matrix inversion (SMI) algorithm in the later stage, and under a high input SNR, the interference suppression performance of the SMI algorithm weakens. In order to suppress the influence brought by noise changes and enhance the interference suppression ability, the DL algorithm is proposed, making the output SINR performance better than that of the SMI.

The core idea of the DL algorithm is to add a diagonal unit matrix to the covariance matrix of the sampled data. By reducing the proportion of interference and noise signals in the covariance matrix of the sampled data, a stable covariance matrix of the sampled data is obtained. On this basis, the optimal weight vector is solved, and then a more robust array pattern function is obtained.

$$\hat{R}_{dl} = \hat{R}_{i+n} + \lambda_{dl} I \tag{4.87}$$

where $\hat{R}_{dl}$ is the covariance matrix of the sampled data after loading, and λ_{dl} is the diagonal loading factor, which is generally an empirical fixed value.

The formula for calculating the weight vector of DL algorithm is expressed as

$$W_{dl} = \frac{\hat{R}_{dl}^{-1} a(\theta_e)}{a^H(\theta_e)\hat{R}_{dl}^{-1} a(\theta_e)} \tag{4.88}$$

Based on the MSINR criterion for solution, the optimal weight vector of DL algorithm can be determined, that is $\min_{w}(1 - \lambda_{dl}) w^{\mathrm{H}} \hat{R}_{dl} w + \lambda_{dl} \|w\|_2^2$, where the constraint conditions can be expressed as

$$w^{\mathrm{H}} \hat{a}(\theta_e) = 1 \tag{4.89}$$

The DL algorithm enhances the stability of the beamforming effect and improves the interference suppression performance. The principle of the algorithm is simple and the operation is easy. However, the specific value of the diagonal loading factor has a significant impact on the beamforming effect of DL algorithm. Both too large and too small diagonal loading factors will seriously affect the degree of interference suppression. Therefore, it is of great significance how to determine the value of the diagonal loading factor according to different expected signal directions and interference conditions.

The principle of DL algorithm is to add data to the covariance matrix of the sampled data. The eigenvalues of the covariance matrix can be introduced into the value range constraint of the loading factor λ_{dl} to avoid the problem of weakened beamforming performance caused by the unreasonable selection of λ_{dl}. The value range is defined as

$$\sigma_{std}[diag(R_{i+n})] < \lambda_{dl} < \frac{trace(R_{i+n})}{M} \tag{4.90}$$

where σ_{std} represents finding the standard deviation, and diag() represents obtaining the elements on the diagonal of the matrix.

According to the determined value range of the loading factor, the intelligent optimization algorithm can be used to assist DL algorithm. The specific value of the loading factor is optimized within the value range by using the intelligent optimization algorithm to determine an appropriate diagonal loading factor value.

One of the major purposes of IRAB is to suppress interference. During the process of global optimization using intelligent optimization algorithms, the optimization criterion of the lowest null in the interference direction can be set. According to the principle of the array pattern, the side lobe levels of the interference positions in the array output pattern are calculated. The sum of the side lobe levels of the interference positions in DL algorithm is taken as the evaluation function for algorithm iteration. Obtain the updated optimal diagonal loading factor. Its fitness function can be expressed as

$$fitness = \min\left\{\sum_{i=1}^{I} sidelobe(\theta_i)\right\} \tag{4.91}$$

where θ_i represents the position of the interference angle.

The steps of UDL algorithm are as follows:

(1) R_{i+n} is estimated based on the sampling data;
(2) Perform eigenvalue decomposition on the covariance matrix and calculate the lower and upper limits of the loading factor.
(3) Take the value range of the loading factor as the search space, and take the sum of the side lobe levels of the interference signal positions in beamforming as the evaluation function;
(4) Iterate the intelligent optimization algorithm, update the optimal diagonal loading factor, and finally use the obtained optimal loading value for diagonal loading.

As shown in Fig. 4.29, the comparison of the UDL algorithm effects and the traditional DL algorithm effects. The simulation adopts a uniform linear array of 20 antenna elements, with a SNR of 10 dB and an interference to noise ratio (INR) of 20 dB. Three signals are randomly generated, and the incident angles are selected

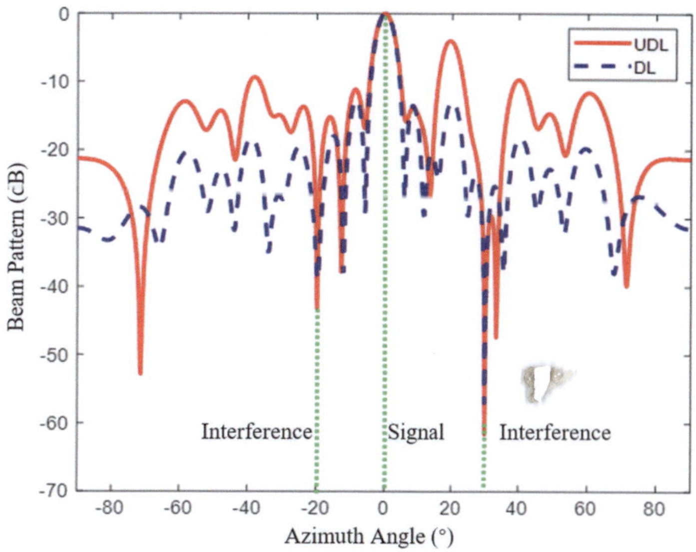

Fig. 4.29 Comparison of beamforming effects between updating diagonal loading factor and fixed-value diagonal loading factor

as 0° (expected signal) and − 20°, 30° (interference signal), respectively. Under the condition that the performance such as the main beam orientation and the lobe width remains unchanged, the suppression degree of UDL algorithm in the − 20° direction is 4.2 dB stronger than that of the DL algorithm, and the suppression degree in the 30° direction is 4.4 dB stronger than that of the DL algorithm. It can be said that the interference suppression effect of the UDL algorithm is better than that of the DL algorithm.

4.3.2.3 Interference Suppression Beamforming Algorithm for Sparse Circular Arrays Based on Joint UDL and CPSO Algorithms

At present, the DL algorithm is mostly applied to uniform linear arrays, and there are few studies on the DL algorithm of conformal sparse arrays. In this study, by using the CPSO algorithm, a sparse circular array interference suppression beamforming algorithm combining UDL and CPSO algorithms is proposed to achieve null suppression in the interference azimuth direction while ensuring the unchanged effect of the main beam in the expected direction, thereby improving the communication quality.

Firstly, based on the SNR, the sparse circular array flow pattern A_s, the expected direction (φ_e, θ_e), and I interference signal directions (φ_i, θ_i) $(i = 1, 2, \ldots I)$, obtain the array received signal model $x_s(t)$, the expected received signal model $r_s(t)$ and the interference signal received signal model $i_s(t)$

$$x_s(t) = A_s s(t) + n(t) = a_s(\varphi_e, \theta_e)e(t) + \sum_{i=1}^{I} a_{si}(\varphi_i, \theta_e)i_i(t) + n(t) \tag{4.92}$$

$$r_s(t) = a_s(\varphi_e, \theta_e)e(t) \tag{4.93}$$

$$i_s(t) = \sum_{i=1}^{I} a_{si}(\varphi_i, \theta_e)i_i(t) \tag{4.94}$$

Calculate the covariance matrix R_{i+n} of noise and interference

$$R_{i+n} = R_i + R_n \tag{4.95}$$

Calculate the value range of λ_{dl} based on the covariance matrix of interference and noise and take it as the search space of the algorithm

$$\sigma_{std}[diag(R_{i+n})] < \lambda_{dl} < \frac{trace(R_{i+n})}{M} \tag{4.96}$$

The sum of the side lobe levels at the interference positions of diagonally loaded beamforming is taken as the evaluation function:

$$fitness = \min\left\{\sum_{i=1}^{I} sidelobe(\theta_i)\right\} \tag{4.97}$$

Finally, CPSO is used for optimization.

4.3.3 Simulation Results of Interference Suppression Beamforming for Conformal Sparse Arrays

4.3.3.1 Simulation Results of Interference Suppression Beamforming Algorithm for Sparse Circular Array Based on Joint UDL and CPSO Algorithms

The simulation adopts a circular array antenna with 25 antenna elements. The number of array elements after sparsity is 12, and the sparsity rate is 44%. The layout of the sparse circular array is maintained as shown in Fig. 4.30. Conduct simulation tests on the azimuth adaptive DL algorithm of sparse circular arrays based on CPSO.

The SNR of the simulation test is 10 dB, and the INR is 20 dB. Three signals are randomly generated, and the incident angles are selected as (180°, 45°) (expected signals) and (120°, 45°), (240°, 45°) (interference signals in the same channel). The azimuth beamforming effects of the sparse circular array and the non-sparse circular

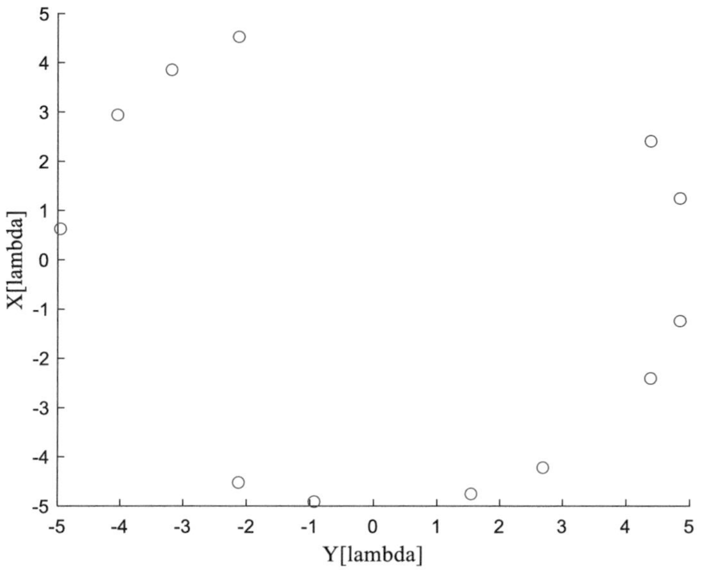

Fig. 4.30 Schematic diagram of the layout of sparse circular wave elements

array are shown in Fig. 4.31. In terms of the comparison of the main beam orientation and main beam width of the array, there is no significant difference in performance between the sparse array and the non-sparse array. In terms of interference signal suppression, the suppression performance of the sparse array is 10 dB (120°, 45°) and 8 dB (240°, 45°) weaker than that of the non-sparse array respectively, but both can achieve good interference suppression effects. On the premise of keeping the gain of the main beam unchanged and sacrificing a certain suppression depth, compared with non-sparse arrays, sparse arrays can still achieve good suppression of interference and reduce the design complexity and cost of array antennas at the same time.

Gradually increase the number of interference signals, test the main beam orientation and beam width of the array, as well as the suppression effect on the interference signals. When the number of interference signals reaches three, four and five, the directions of interference signals are respectively (60°, 45°), (120°, 45°), (240°, 45°); (60°, 45°), (120°, 45°), (240°, 45°), (300°, 45°); (30°, 45°), (60°, 45°), (120°, 45°) and (240°, 45°), (300°, 45°); With other conditions remaining unchanged, the formation effect of the sparse circular array is shown in Figs. 4.32, 4.33 and 4.34.

It is observed that as the number of interference signals gradually increases, after the sparse array is diagonally loaded, the performance of the main beam does not change significantly, and the suppression effect on the interference signals is gradually weakening, but interference suppression can still be achieved.

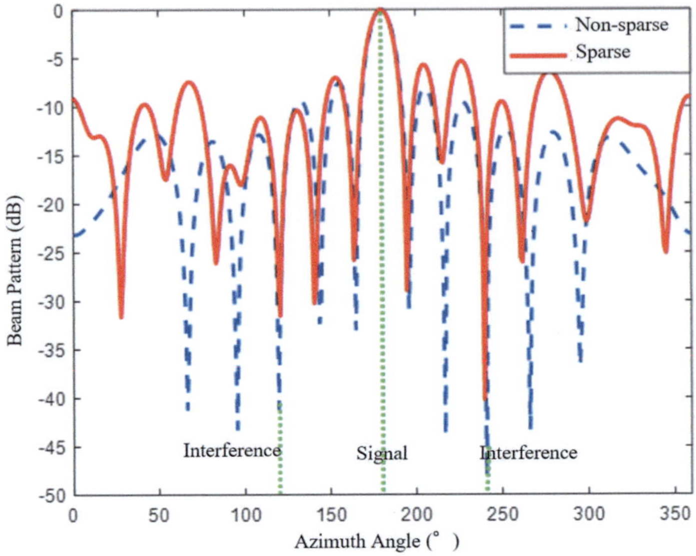

Fig. 4.31 Comparison of beamforming effects between sparse and non-sparse arrays

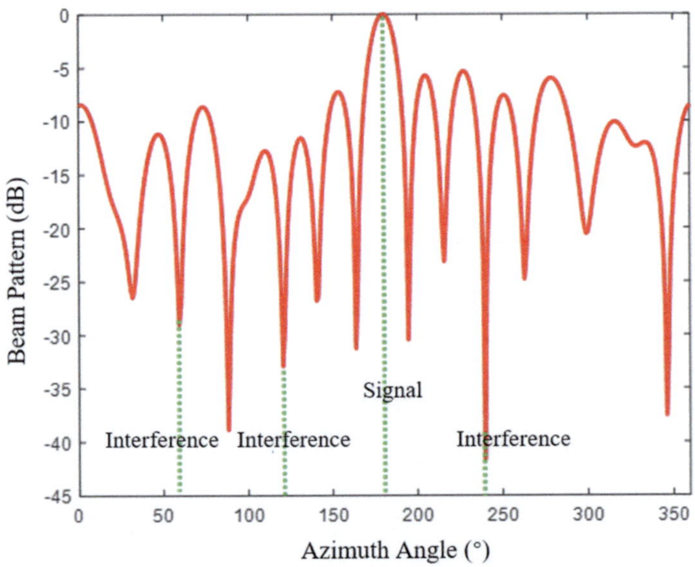

Fig. 4.32 The beamforming effect of sparse arrays under three interferences

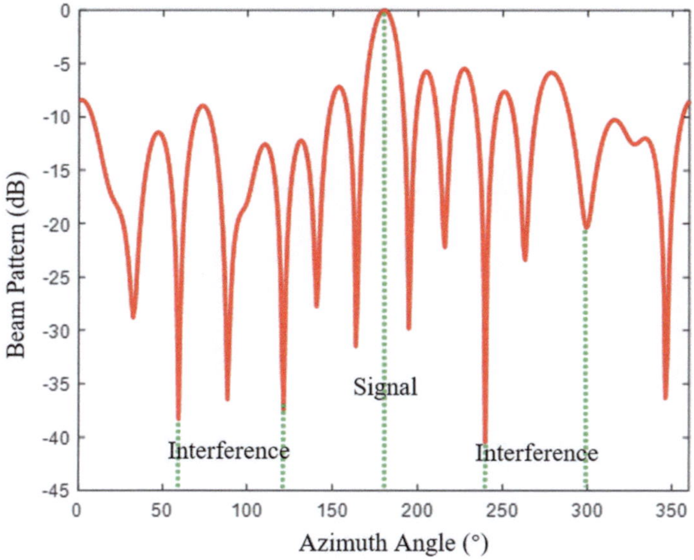

Fig. 4.33 The beamforming effect of sparse arrays under four interferences

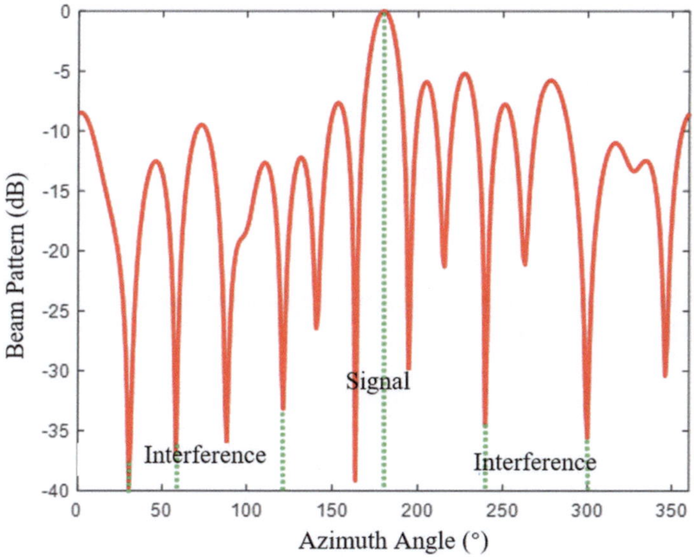

Fig. 4.34 The beamforming effect of sparse arrays under five interferences

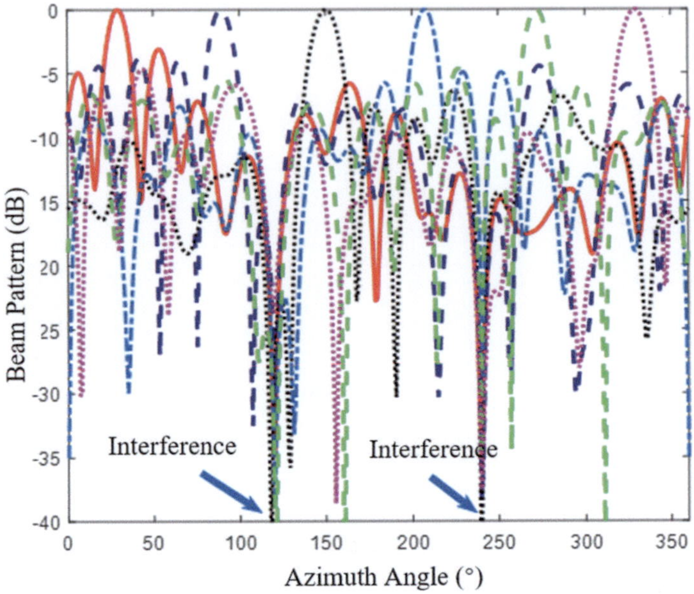

Fig. 4.35 The beam scanning capability of sparse circular arrays

Figure 4.35 shows the azimuth beam scanning performance of the sparse circular array after using the beamforming algorithm proposed in this study. It can still achieve 360° omnidirectional scanning in the azimuth direction, and the scanning beam width remains unchanged, still having the omnidirectional performance of the circular array.

4.3.3.2 Simulation Results of Interference Suppression Under Multi-user Access

This study designs and verifies the suppression degree of interference and the SINR of each user by the proposed beamforming algorithm when accessing multiple users. The design idea is roughly as follows: Each user is regarded as the direction of the incoming signal wave, and other users are regarded as interference signals relative to this user. The sparse circular array interference suppression beamforming algorithm combining UDL and CPSO algorithms is applied to test the SINR of each user under the access of 6 users.

The simulation conditions are as follows: There are a total of 6 signal directions, namely (30°, 45°), (60°, 45°), (90°, 45°), (120°, 45°), (150°, 45°), and (180°, 45°). The SNR is 26 dB for all of them, and the number of circular wave elements is 25. The beam pattern for concurrent communication among six users is shown in Fig. 4.36.

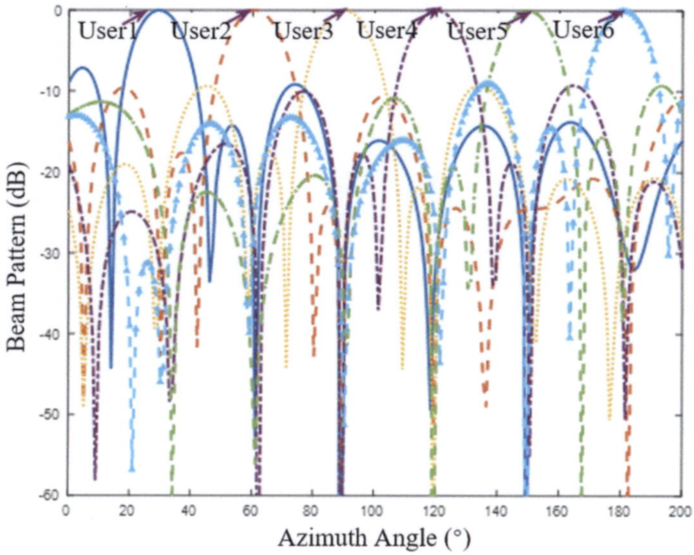

Fig. 4.36 Six-user signal beamforming effect

In order to study the interference situation of other user signals on the expected user signal, this study conducts the simulation as shown in Fig. 4.37. The direction maps of the five interfering users other than the expected user are superimposed. According to the calculation such as beam width, the array implemented by beamforming is above 20 dB. In the expected azimuth direction of 30°, the interference suppression degree reaches 25 dB in the expected user direction. In the expected azimuth direction of 60°, the interference suppression degree reaches 24 dB in the expected user direction. In the expected azimuth direction of 90°, the interference suppression degree at the expected user direction reaches 21 dB. In the expected azimuth direction of 120°, the interference suppression degree at the expected user direction reaches 21.6 dB. In the expected azimuth direction of 150°, the interference suppression degree at the expected user direction reaches 21.6 dB. In the expected azimuth direction of 180°, the interference suppression degree reaches 20 dB in the expected user direction. When the SINR in all directions is above 20 dB, etc., the suppression of interference from other users under multi-user reception has been achieved, and it has the effect of space division multiplexing.

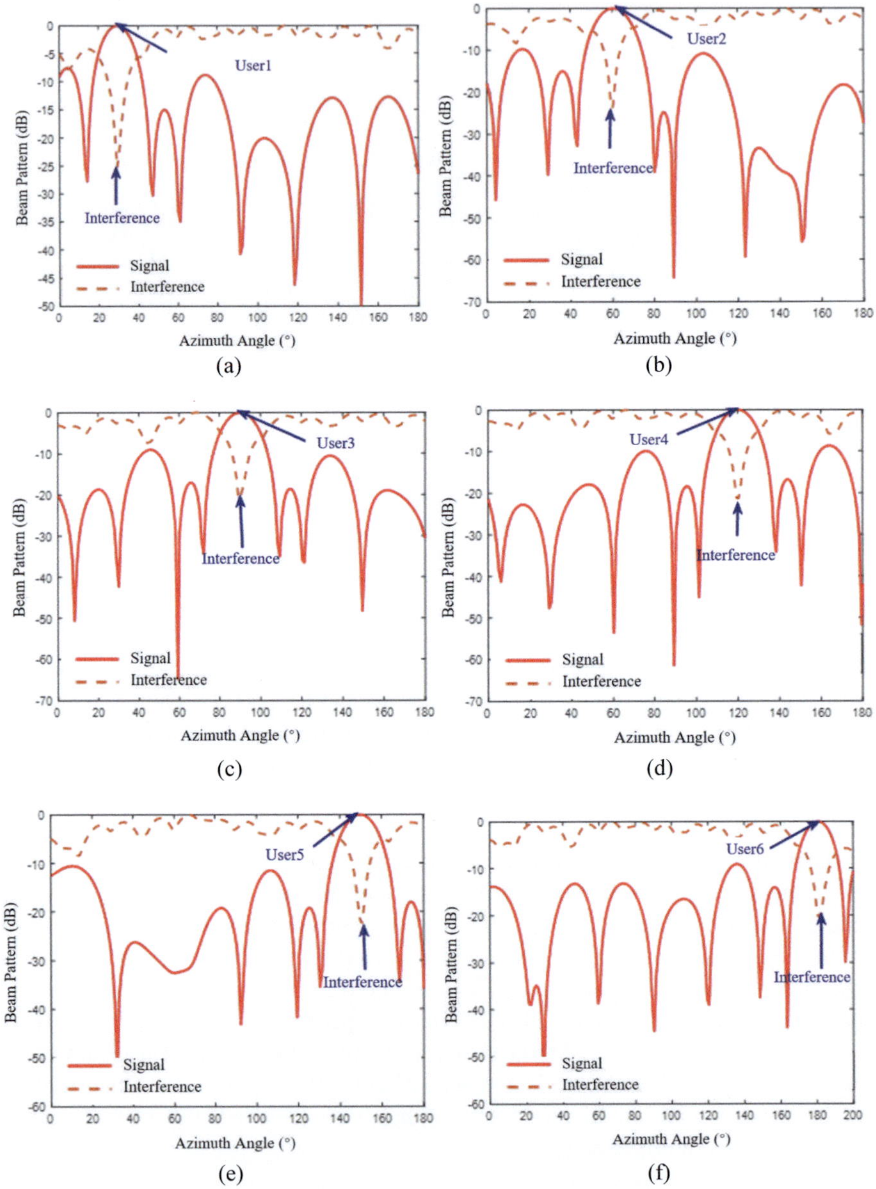

Fig. 4.37 Interference suppression effect at **a** 30°; **b** 60°; **c** 90°; **d** 120°; **e** 150°; **f** 180°

References

1. E. J. Candes, J. Romberg and T. Tao, "Robust uncertainty principles: Exact signal reconstruction from highly incomplete frequency information," *IEEE Transactions on information theory*, vol.52, no. 2, pp. 489-509, 2006.
2. J. Grover, A. Jain, and N. S. Chaudhari, "Unmanned aerial vehicles operated emergency ad hoc networks," *2017 7th International Conference on Communication Systems and Network Technologies (CSNT)*, Nagpur, India, 2017, pp. 92–96.
3. X. Liu and N. Ansari, "Resource Allocation in UAV-Assisted M2M Communications for Disaster Rescue," *IEEE Wirel. Commun. Lett.*, vol. 8, no. 2, pp. 580–583, Apr. 2019.
4. L. Josefsson and P. Persson. *Conformal array antenna theory and design*. John wiley & sons, USA, 2006.
5. Z. C. Hao, M. He, and W. Hong, "Design of a millimeter-wave high angle selectivity shaped-beam conformal array antenna using hybrid genetic/space mapping method," *IEEE Antennas Wireless Propag., Lett.*, vol. 15, pp. 1208–1212, Nov. 2015.
6. J.-L. Guo and J.-Y. Li, "Pattern synthesis of conformal array antenna in the presence of platform using differential evolution algorithm," *IEEE Trans. Antennas Propag.*, vol. 57, no. 9, pp. 2615–2621, Sep. 2009.
7. Y.-Y. Bai, S. Xiao, C. Liu, and B.-Z. Wang, "A hybrid IWO/PSO algorithm for pattern synthesis of conformal phased arrays," *IEEE Trans. Antennas Propag.*, vol. 61, no. 4, pp. 2328–2332, Apr. 2012.
8. K. Owoola, T. Xia, A. Wang, A. Umar, and R. G. Akindele, "Pattern Synthesis of Uniform and Sparse Linear Antenna Array Using Mayfly Algorithm," *IEEE Access*. vol. 9, pp. 77954–77975, 2021.
9. Q. Ling, C. Yang, Y. Yu, T. Liu, and X. Sheng, "Sparsity and Adaptivity for Sensor Location Optimization of Underwater Conformal Vector Array," *Global Oceans 2020: Singapore – U.S. Gulf Coast, Biloxi*, MS, USA, 2020, pp. 1–5.
10. MAN Fakher Feb. 2015 Pattern synthesis for a conformal array antenna mounted on a paraboloid reflector using genetic algorithms Appl. Comput. Electromagn. Soc. 30 2 230 236
11. M. Fu, Z. Zheng, and W.-Q. Wang, "two-dimensional DOA Estimation for Nested Conformal Arrays via Sparse Reconstruction," *IEEE Commun. Lett.*, vol. 25, no. 3, pp. 980-984, Mar. 2021.
12. H. Zheng, C. Zhou, Z. Shi, Y. Gu, and Y. D. Zhang, "Coarray Tensor Direction-of-Arrival Estimation," *IEEE Trans. Signal Process.*, vol. 71, pp. 1128-1142, Mar. 2023.
13. H. Zheng, C. Zhou, Z. Shi, and G. Liao, "Sub-Nyquist Tensor Beamformer: A Coprimality Constrained Design," *IEEE Trans. Signal Process.*, vol. 71, pp. 4163-4177, Aug. 2023.
14. L. Liang, C. Jin, H. Li, J. Liu, Y. Jiang, and J. Zhou, "A hybrid algorithm of orthogonal perturbation method and convex optimization for beamforming of sparse antenna array," *Electromagnetics*, vol. 40, no. 4, pp. 227–243, Apr. 2020.
15. M. Lv, J. Zhao, and J. Xu, "Pattern Synthesis for Sparse Conformal Arrays Using Improved Grey Wolf Optimizer," *2021 International Conference on Microwave and Millimeter Wave Technology (ICMMT)*, Nanjing, China, 2021, pp. 1–3.
16. G. Oliveri, E. T. Bekele, F. Robol, and A. Massa, "Sparsening Conformal Arrays Through a Versatile BCS-Based Method," *IEEE Trans. Antennas Propag.*, vol. 62, no. 4, pp. 1681–1689, Apr. 2014.
17. Y. Liu, Z. Nie, and Q. H. Liu, "Reducing the Number of Elements in a Linear Antenna Array by the Matrix Pencil Method," *IEEE Trans. Antennas Propag.*, vol. 56, no. 9, pp. 2955-2962, Sep. 2008.
18. Y. Liu, Q. H. Liu, and Z. Nie, "Reducing the Number of Elements in Multiple-Pattern Linear Arrays by the Extended Matrix Pencil Methods," *IEEE Trans. Antennas Propag.*, vol. 62, no. 2, pp. 652-660, Feb. 2014.
19. F. Viani, G. Oliveri, and A. Massa, "Compressive Sensing Pattern Matching Techniques for Synthesizing Planar Sparse Arrays," *IEEE Trans. Antennas Propag.*, vol. 61, no. 9, pp. 4577-4587, Sep. 2013.

20. Y. B. Tian and J. Qian, "Improve the performance of a linear array by changing the spaces among array elements in terms of genetic algorithm," *IEEE Trans. Antennas Propag.*, vol. 53, no. 7, pp. 2226–2230, Jul. 2005.
21. G. Liu, H. Zhu, K. Wang, J. Mou, P. Zheng, and G. Wei, "Sparse Conformal Array Synthesis Based on Multiagent Genetic Algorithm," *2022 IEEE 10th Asia-Pacific Conference on Antennas and Propagation (APCAP)*, Xiamen, China, 2022, pp. 1–2.
22. M. M. S. Taheri, A. R. Mallahzadeh, and A. Foudazi, "Shaped beam synthesis for shaped reflector antenna using PSO algorithm," *in Proc. Eur. Conf. Antennas Propag.*, pp. 1–5, Mar. 2012.
23. H. Li, Y. Jiang, Y. Ding, J. Tan, and J. Zhou, "Low-Sidelobe Pattern Synthesis for Sparse Conformal Arrays Based on PSO-SOCP Optimization," *IEEE Access*, vol. 6, pp. 77429–77439, Nov. 2018.
24. N. Nagendra and L. Pappula, "Evolution of conformal antenna design with the aid of probability improved crow search algorithm," *Data Technol. Appl.*, vol. 55, no. 3, pp. 353–371, Jun. 2021.
25. A. Hashim and Fatma, "Snake optimization: A novel meta-heuristic optimization algorithm," *Knowledge-Based Syst.*, vol. 235, pp. 108320, Apr. 2022.
26. Z. Bao and J. Yu, "Thinning Curved Surface Arrays: Application of the Genetic Algorithm (GA)," *IET International Radar Conference*, Hangzhou, China, 2015, pp. 1–4.
27. I. M. Sobol, D. I. Asotsky, A. Kreinin, and S. S. Kucherenko, "Construction and Comparison of High-Dimensional Sobol' Generators," *Wilmott*, vol. 2011, no. 56, pp. 64-79, Nov. 2011.
28. S. Zhao, P. Wang, A. A. Heidari, X. Zhao, C. Ma, and H. Chen, "An enhanced Cauchy mutation grasshopper optimization with trigonometric substitution: engineering design and feature selection," *Eng. Comput.*, vol. 38, no. 5, pp. 4583-4616, Dec. 2022.
29. S. M. B. Seyed and S. Y. Samaneh, "IWOA: An improved whale optimization algorithm for optimization problems," *J. Comput. Des. Eng.*, vol. 6, no. 3, pp. 243–259, Jul. 2019.

Chapter 5
Conclusion

Compared to planar array antennas, conformal antennas offer superior conformability to carrier surfaces. When deployed on aircrafts (e.g., balloons or UAVs), they maintain excellent aerodynamic characteristics. Furthermore, conformal array antennas can achieve broad coverage (e.g., half-space beam coverage) of directional beams via incorporating phase shifting networks or switching circuits.

5.1 Contributions of This Book

New methodologies and novel designs have been proposed in this book. The primary contributions of this book are as follows.

- Based on the analysis and performance comparison of conformal strip-dipole and bowtie-dipole, a conformal log-periodic balloon antenna has been proposed in this paper. Employing the conformal bowtie-dipoles as its elements, the balloon antenna has a lot of advantages such as small size, wide bandwidth, stable endfire patterns and reasonable radiation efficiency and gain. Measured results of our prototype antennas fabricated with aluminum foils and PVC balloons have demonstrated the good radiation performance of the proposed design and its potential of application in VHF-band wireless communication systems.
- A balloon-based conformal array antenna with high gain, broadened bandwidth, and steerable directional beam is proposed for emergency communication applications in low-range hilly forest environments. The basic array element is a cylindrically conformal microstrip antenna with broadened bandwidth. A 1×4 conformal subarray is then designed for beam sharpening in the elevation plane. To further improve the gain and enable beam steering for full azimuthal coverage, the balloon is surrounded by multiple conformal subarrays and fed by a specially designed conformal flexible phase shifting network. Finally, a prototype is fabricated and

X. Gao et al., *Conformal Array Antennas and Beamforming Technologies for Emergency Communication Applications*, Modern Antenna,
https://doi.org/10.1007/978-981-95-2905-6_5

measured to validate the effectiveness of the designs. Experimental results show that the designed system exhibits a high gain of around 15 dB and a bandwidth of 74 MHz with a steerable directional beam along the azimuthal plane at the 2.4 GHz band, which agrees well with the simulation. The antenna can potentially be applied to emergency communications in the low-range hilly forest environment.

- A balloon-based conformal heteroge neous array antenna with ultrawide bandwidth and half-space beam coverage is proposed for emergency communication applications. Consisting of nine conformal broadside and endfire subar rays, half-space beam coverage can be achieved via 1-D full-azimuthal beam switching and a combination of the radiation patterns. To widen the bandwidth and enhance the gain, the broadside subarray employs 1 × 4 dumbbell shaped full-wavelength dipoles cross-fed in series and backed by a miniaturized AMC; the endfire subarray utilizes 2 × 3 size-tapered half-wavelength dipoles with optimized spacing to achieve the desired radiation characteristics. Finally, the prototypes of two conformal subarrays were fabricated and measured. Experimental results show the proposed antenna exhibits an overlapped bandwidth of 720 MHz and 360° × 108° beam coverage at the 2.4-GHz band, which agree well with simulation. The superior radiation performance of the proposed conformal heterogeneous array has demonstrated its huge potential in emergency communication applications.
- A quad-polarization reconfigurable conformal array of wide coverage range and high gain is presented for UAV communication systems. The proposed single-port excited miniaturized element utilizes a low-complicity reconfigurable feeding network to realize two orthogonal LPs, LHCP, and RHCP polarization modes. A quad-polarization reconfigurable UAV conformal sparse array is proposed, designed, and characterized. The measurement results show that the 3 dB AR bandwidth is around 1.3% for the CP mode, and the − 10 dB impedance bandwidths are 7.7% and 7.3% for the CP and LP modes. The measured gain of the conformal array is over 14 dBi for all polarization modes. A relatively wide beam coverage range of 50° × 180° is achieved. In summary, the good radiating performance of the proposed conformal sparse array antenna makes it potentially be applied to the UAVs with payload constraints for emergency communication applications.
- A helmet-based hemisphericalconformal wearable array antenna with broad bandwidth, wide-angle coverage and reconfigurable dual CPs is proposed for WBAN applications. By utilizing a single-feed polarization reconfigurable crossed dipole incorporated with parasitic patches above a ground, a broadband conformal radiating element with unidirectional radiations is designed. The element integrated with PIN-diodes can flexibly operate in two orthogonal CP modes by switching the on–off states of the diodes. To broaden the operating bandwidth, four sectorshaped patches working as parasitic radiators are loaded on four quadrants of the crossed dipole. Besides, a 12-element helmet-based hemispherical-conformal wearable array antenna with two reconfigurable CP modes is designed, which is arranged in special configuration for improving the gain, coverage range and orientation flexibility for WBAN systems. The experimental results show that the proposed

hemispherical-conformal wearable array exhibits an overlapped 3-dB ARBW of 29.6% and -10-dB IMBW of 69.8% at around 3 GHz for both reconfigurable CP modes. Besides, a peak realized gain more than 8 dBi and a beam coverage range of $110^\circ \times 360^\circ$ are obtained. The superior radiating performance of the proposed antenna has proved its huge potential for wearable WBAN applications.

- A L1-SVD based one-dimensional coherent signal DOA estimation algorithm is proposed to solve the difficulty of DOA estimation caused by the rank reduction of the covariance matrix of coherent signals, which is suitable for uniform circular arrays and sparse circular arrays. A two-dimensional coherent signal DOA estimation algorithm based on conformal circular array is also proposed to solve the problem of large calculation and difficulty in estimating two-dimensional coherent signal DOAestimation. Then, a conformal sparse optimization method based on ISO is proposed to solve the problems of excessive array elements caused by large conformal array size and poor beam radiation performance caused by high maximum sidelobe level. Finally, an interference suppression beamforming method for conformal sparse arrays based on joint UDL and CPSO is proposed to solve the problems of serious interference and difficulty in improving communication capacity when multi-user concurrent transmission in the same channe.

5.2 Future Work

This book focuses on the design of conformal array antennas and beamforming technologies for emergency communication applications. Several novel designs and algorithms are proposed. Meanwhile, a variety of open research topics remain.

- The gain of the proposed conformal array antennas is below 15 dBi. Therefore, conformal array antennas with higher gain need to be developed for ultra-long-distance wireless communications (e.g., a gain of higher than 20 dBi for satellite communication).
- The beamforming technologies developed in this book remain theoretical. Integrating them with practical antenna designs will represent a significant advancement.
- The conformal shapes addressed in this book are limited to cylindrical and spherical surfaces. Designs and corresponding algorithms for conformal antennas on more complex geometries or arbitrary curved surfaces would be benificial for more complicated scenarios.

MIX
Papier aus verantwortungsvollen Quellen
Paper from responsible sources
FSC® C105338

If you have any concerns about our products,
you can contact us on
ProductSafety@springernature.com

In case Publisher is established outside the EU,
the EU authorized representative is:
Springer Nature Customer Service Center GmbH
Europaplatz 3, 69115 Heidelberg, Germany

Printed by Libri Plureos GmbH
in Hamburg, Germany